Calculating the Weather

This is Volume 60 in the
INTERNATIONAL GEOPHYSICS SERIES
A series of monographs and textbooks
Edited by RENATA DMOWSKA and JAMES R. HOLTON

A complete list of books in this series appears at the end of this volume.

Calculating the Weather

Meteorology in the 20th Century

Frederik Nebeker

IEEE CENTER FOR THE HISTORY OF
ELECTRICAL ENGINEERING
RUTGERS UNIVERSITY
NEW BRUNSWICK, NEW JERSEY

ACADEMIC PRESS
San Diego New York Boston London
Sydney Tokyo Toronto

Front cover photograph: NOAA-8 visual imagery of Hurricane Gloria, September 25, 1985, at 12:39 GMT. The islands of Cuba, Hispaniola, and Puerto Rico are faintly visible near the bottom of the print. Courtesy of National Oceanic and Atmospheric Administration; National Environmental Satellite, Data, and Information Service; National Climatic Data Center; and Satellite Data Services Division.

This book is printed on acid-free paper. ♾

Academic Press, Inc.
A Division of Harcourt Brace & Company
525 B Street, Suite 1900, San Diego, California 92101-4495

United Kingdom Edition published by
Academic Press Limited
24-28 Oval Road, London NW1 7DX

Library of Congress Cataloging-in-Publication Data

Nebeker, Frederik.
Calculating the weather : meteorology in the 20th century / Frederik Nebeker.
p. cm. -- (International geophysics series : v. 60)
Includes bibliographical references and index.
ISBN 0-12-515175-6
1. Meteorology—Methodology. I. Title. II. Series.
QC871.N344 1995
551.5'028--dc20 94-37625
CIP

PRINTED IN THE UNITED STATES OF AMERICA
95 96 97 98 99 00 QW 9 8 7 6 5 4 3 2 1

Contents

Part II Meteorology in the First Half of the 20th Century

Chapter 5 Vilhelm Bjerknes's Program to Unify Meteorology

Chapter 6 Lewis Fry Richardson: The First Person to Compute the Weather

Chapter 7 The Growth of Meteorology

Chapter 8 Meteorological Calculation in the Interwar Period

Chapter 9 The Effect of World War II on Meteorology

Part III The Beginning of the Computer Era in Meteorology

Chapter 10 John von Neumann's Meteorology Project

Chapter 11 The Acceptance of Numerical Meteorology

Chapter 12 The Unification of Meteorology

Chapter 1 | Introduction

Three Traditions in Meteorology

Recording weather observations, explaining the action of the atmosphere, and predicting wind and rain are all ancient practices. The Babylonians did all three some 3000 years ago. The Greeks kept records of wind direction from the time of Meton (*ca.* 430 B.C.), had a theoretical meteorology from the time of Aristotle (*ca.* 340 B.C.), and were advised by "weather signs" from time immemorial. In the 17th century new instruments, such as the thermometer and barometer, permitted the measurement of elements of the weather; René Descartes, Edmond Halley, and others speculated on the causes of winds; and almanacs made weather prognostications widely available. Through to the 20th century these three activities have been the principal ways people have manifested scientific interest in the weather: an empirical activity of making records of observations and then trying to infer something from these records, a theoretical activity of explaining atmospheric phenomena on the basis of general principles, and a practical activity of predicting the weather.

The activities of observer, natural philosopher, and forecaster were of course related, and the term "meteorology" has always encompassed all three. However, in the course of the 19th century, as the number of people doing meteorology increased, the empirical, theoretical, and practical activities became more distinct. Many of those working in the empirical tradition made the average weather their principal interest as they cultivated a descriptive science—called "climatology" from mid-century on—based on weather statistics. Many of those working in the theoretical tradition made the laws of physics their starting point and established, as a branch of the science, dynamical meteorology. Weather forecasting became a profession with the initiation of daily forecasting by national meteorological services in the 1870s and thereafter. Yet the work of the empiricists—some of whom derided "theorizing"—involved little physics. The theorists, for their part, seldom drew on the vast store of meteorological observations in composing their treatises. And the forecasters based their predictions on only a small amount of data and hardly any theory at all, hence their work was regarded by many empiricists and theoreticians as unscientific.

The three traditions continued their separate developments until the middle of the 20th century.[1] Then rather suddenly the connections between them became stronger and more numerous, and meteorologists talked frequently about a unification of meteorology. This unification, although it depended on the new availability of electronic computers in the 1950s and 1960s, was the culmination of a transformation of the science that began much earlier.

The Unification of Meteorology

In 1903 Vilhelm Bjerknes, a Norwegian physicist-turned-meteorologist began advocating a calculational approach to weather forecasting, believing it possible to bring together the full range of observation and the full range of theory to predict the weather. Bjerknes's program, which if successful would have united the three traditions, gained the attention and applause of meteorologists everywhere, but progress was slow.

The first person to make a full trial of Bjerknes's program was the English scientist Lewis Fry Richardson. While working as a scientist in industry, Richardson discovered an arithmetical method of solving partial differential equations. It seems that he turned to meteorology because he thought that he could there apply his method with success. He devised, during and shortly after World War I, an algorithmic scheme of weather prediction based on the method. This scheme required certain types of data and certain types of theories, and the inappropriateness of much of what was then available motivated Richardson to develop new observational techniques and new theories. He was motivated also to do a lot of what is now known as numerical analysis. Richardson tested the scheme, taking 6 weeks to calculate a 6-hour advance in the weather. The results were egregious. Richardson's work, which was widely noticed, convinced meteorologists that a computational approach to weather prediction was completely impractical.

In the interwar period meteorology became established throughout the Western world as an academic discipline and as a full-fledged profession. Shortly after World War I a group of meteorologists under Bjerknes's direction in Bergen introduced the concepts of cold and warm fronts, the polar front, and air masses, all of which proved to be useful in forecasting. Although the Bergen techniques were largely independent of dynamical meteorology, the latter did begin to be quite useful to forecasters just before World War II, especially through the work of Carl-Gustaf Rossby. Rossby derived an equation giving the speed of certain long-wavelength waves in westerly wind currents, and he showed how the assumption of constant vorticity of winds could be used to calculate air movement. Several calculating aids were devised to make it easier for forecasters to use Rossby's results. Also, in the interwar period meteorologists began using punched-card machines and vigorously pursued the search for weather cycles, equipped with a panoply of special-purpose calculating aids.

During World War II meteorology came to be perceived as having great military value, and this fact had great effects on the science. Along with a great increase in the number of meteorologists, there were important theoretical and instrumental advances, and after the war governmental support of meteorology remained far above prewar levels. During the war, meteorologists became more interested in objective methods of forecasting and, aided by punched-card technology, showed the great practical value of climatology.

The first electronic computer, the ENIAC, was completed just as World War II ended, and at that time John von Neumann began making plans to build, at the Institute for Advanced Study in Princeton, a much more powerful and versatile machine devoted to the advancement of the mathematical sciences. An important objective for von Neumann was to demonstrate, with a particular scientific problem, the revolutionary potential of the computer. He chose for this purpose weather prediction and in 1946 established the Meteorology Project at the Institute. The Project had a slow start and the Institute computer took longer to build than was expected, but by 1956, when the Project ended, von Neumann's expectations had been fulfilled: it had been shown that a physics-based algorithm could be used to predict large-scale atmospheric motions as accurately as human forecasters could, and it had been shown that computer technology could carry out such algorithms fast enough and reliably enough for the forecasts to be useful.

In the 1950s and 1960s the computer became a standard tool in meteorology, and most other calculating aids were abandoned. By 1970 much data handling and data analysis were done by computer, theorists used computer modeling and numerical experimentation as principal modes of investigation, and in the industrialized countries most weather services used computers in making forecasts. Great advances were made in the empirical, theoretical, and practical traditions through the facilitation of computation. The importance of forecasting models gave direction to data gathering and to theorizing, as the observational meteorologists and the theorists often had an eye to the use of their results in such models. Quite generally, climatologists, dynamical meteorologists, and forecasters came to use similar computer models. Indeed, computing power made possible so many new connections between the traditions that they may be said to have merged. At the same time, the use of the computer led to the discovery of so-called "chaotic systems" and thence to the recognition that there may well be fundamental limits to predicting the weather.

Transformations of Meteorology

Thus in the course of the 20th century meteorology became a unified, physics-based, and highly computational science. Many meteorologists have remarked on the great changes the science has undergone. Jule Charney, for example, spoke of a "technological-scientific transformation" (1987, p. 168), and George P. Cressman wrote, "The development of the electronic computer changed everything" (1972, p. 181).

This 20th-century transformation was comparable in import to two earlier transformations. In the second half of the 17th century, meteorological observation changed from description almost entirely qualitative to description largely quantitative, as atmospheric pressure, humidity, precipitation, wind direction, and wind force all came to be measured. Meteorology became less a branch of natural philosophy and more an independent, empirical science, and, although most meteorological explanation remained nonquantitative, the 17th-century transformation did make descriptive meteorology a quantitative science.

A second transformation occurred in the second half of the 19th century with the development of the weather map as the basic tool of meteorological description, analysis, and prognostication. The telegraph made possible the construction of same-day weather maps, and the great increase in commerce made weather and climate information more valuable. This transformation was largely organizational, involving the establishment of national weather services, of networks of observers, and of international cooperation among meteorologists.

Technological advances were vital to both these transformations: the first was based on new instrumentation, and the second owed much to improved means of communication. The 20th-century transformation was even more indebted to new technology. Radio led to great expansions of observational networks, with ship- and buoy-to-shore communication and the transmission of meteorological data from instruments carried aloft by balloons and the new technologies of airplanes, rockets, and satellites. Radar opened a new window on the atmosphere. Most important, however, was calculating technology. The effective use of the vastly increased capacity for observing the weather, the maturation of dynamical meteorology, and the great improvement in forecasting technique were all dependent on new calculating technology, principally the electronic computer. It was, moreover, the computer that made possible many of the new links between the empirical, theoretical, and practical traditions, as well as the links between meteorology and other disciplines such as oceanography, hydrology, glaciology, and aeronautics. So the 20th-century transformation may be described as having made meteorology a computational science.

Any overview of meteorology is liable to slight the diversity of the science. The account that follows focuses on certain lines of development and makes little or no mention of other lines, such as studies of atmospheric chemistry, cloud formation, or atmospheric tides, or of optical, electrical, magnetic, and acoustic properties of the atmosphere. Moreover, meteorology has, to some degree, developed independently in every country, and here national differences are not emphasized.

Algorithms, Calculation, and Computation

Because computation is central to the history of modern meteorology, it may be worthwhile to distinguish some related concepts. An algorithm is a fully specified, step-by-step procedure. The specification usually consists of a list of the operations to be carried out sequentially, although a full specification would include a description of the basic mathematical, logical, or physical operations that appear as steps in the procedure. Examples of algorithms are the set of instructions accompanying a video-cassette recorder, the procedure one learns in high-school geometry for constructing the perpendicular bisector of a line segment, and any computer program.

It is useful to distinguish between calculation and computation. Calculation, the broader concept, may be defined as the carrying out of a quantitative algorithm, that is, the manipulation of quantities according to a stated procedure. Computation, on the other hand, may be defined as the carrying out of an arithmetical algorithm, where the steps of the algorithm involve, besides simple logical operations, only addition, subtraction, multiplication, and division. Computation is what computers do. In a computation the quantities are handled as strings of digits. In a calculation, by contrast, the quantities may be represented as alphabetic symbols, lengths on a slide rule, areas on a graph, or voltages in an electric circuit. Whereas a computation involves only arithmetic, the steps of a calculation may be any symbolic or physical operation, such as differentiation, manipulating a slide rule, or finding the area between two curves. The distinction I draw here between computation and calculation is generally consistent with ordinary usage: according to *Webster's Third New International Dictionary* (1981, p. 315) ". . . CALCULATE is usu[ally] preferred in ref[erence] to more complex, difficult, and lengthy mathematical processes . . . COMPUTE is often used for simpler mathematical processes, esp[ecially] arithmetical ones . . ."[2]

The distinction is important in this historical account. One of its themes is that, in meteorology, computations came to replace other sorts of calculation and that this process was given tremendous impetus by the availability of electronic digital computers in the 1950s and 1960s. Concomitantly, the great variety of calculating aids used by meteorologists—mathematical tables, nomograms and other graphical devices, special-purpose slide rules, computing forms, and analog computers—were replaced by a single general-purpose device. (There were several general-purpose calculating aids before the electronic computer—tables of logarithms, the standard slide rule, the differential analyzer, and punched-card machines—but these were much less powerful than the computer.) Another way of expressing this is to say that a great variety of algorithms, many of which involved the physical representation of quantities as lengths, areas, or voltages, came to be replaced by computer programs.

Meteorologists still do a great deal of "calculating" in using the techniques of mathematical analysis to deduce the consequences of certain laws or assumptions. Here, however, 'calculate' has a different meaning: not the carrying out of an algorithm, but the blazing of a logical trail. Even this sort of calculation, it turns out, is being replaced by computer-implemented algorithms as meteorologists increasingly investigate the consequences of a set of assumptions by numerical experimentation rather than by logical deduction, and the analytic skills of meteorologists have apparently declined as a result. In 1987 Philip Thompson wrote, "Mathematical analysis appears to be a dying or lost art, and I would argue for a better balance between analytical and numerical methods" (p. 636).

Forces Leading to an Increased Use of Algorithms

In 19th-century meteorology—empirical, theoretical, and practical—calculation had only secondary roles. One type of calculating aid, numerical tables, was extensively used, especially for what was called data reduction, which involved converting units of measure, making corrections to instrumental readings, and computing quantities measured indirectly. In the early decades of the 20th century, meteorologists and theorists as well as empiricists and forecasters came to use a great many other calculating aids, such as nomograms, plotting forms, special-purpose slide rules, and computing forms.

Theorists made use of calculating aids because as models of atmospheric phenomena became more mathematical, calculating techniques became more important in deducing the behavior of the models. Conversely, the existence of more effective calculating techniques made the mathematical specification of a theory more useful: there was more reason to specify, mathematically and fully, a hydrodynamic or thermodynamic model when it was possible to get numerical predictions as a result. Related to the increasing importance of mathematical modeling was the establishment of numerical experimentation as a principal methodology, since with a fully specified model one can, provided the calculations do not take too long, carry out controlled experiments.

There was in fact a steady movement toward the increased use of algorithms and therefore toward the increased use of calculating aids. Meteorology is hardly unique in this respect. In recent decades many activities in many branches of science have become algorithmic: data are gathered and processed by computer, theoretical models are implemented on computers and their behavior is investigated by numerical experiments, and statistical algorithms are used to discover correlations and other patterns in data and to measure degree of fit between data and model. Indeed, the common attitude in many sciences is that an explanation of a phenomenon is incomplete unless it is so fully specified that it allows the simulation of the phenomenon on a computer. Thus, the work of scientists has increasingly become the devising of algorithms. In meteorology many factors contributed to the movement toward increased use of algorithms, but the main driving forces were what I call "data push," "theory pull," and the attraction of "science-not-art."

It was mainly the climatologists and other empiricists who were impelled by data push, the desire to make something of the large and ever-increasing store of data. When there are few data, one can deal with them in many ways. But when there are a great many, systematization and even automation may be necessary if all the data are to be dealt with. For example, in the interwar period, as the flood of meteorological data swelled, national weather services in many countries began using punched-card machines simply to be able to process in the most basic ways—mainly tabulating and averaging—the reports of observations coming in from ships, airplanes, and land stations. Quite generally, efforts to find regularities in the data, especially by the use of statistical techniques, frequently involved calculating aids. So algorithms became increasing important to the empirical meteorologists.

It was mainly the practitioners of dynamical meteorology who felt theory pull, the desire to connect theory to measurable phenomena. Usually such connections involved extensive calculation, since the theoretically derived formulas that described particular physical processes could seldom be immediately applied to the welter of events going on in the atmosphere. Typically a great deal of work, both with the theoretical formulas and with the data, had to be carried out before any correspondence between theory and measurement was apparent. Since most of this work was calculation, algorithms became increasingly important to the theoretical meteorologists.

The operative metaphor may bear some elaboration. Meteorological measurements are piled on the ground, and meteorological theory is situated somewhere above. There are two ways connections are established: either the data, through the medium of meteorologists, push their way up to general statements or the general statements pull on the medium of meteorologists to make connections with relevant data. It is the meteorologists who are pushed and pulled, and what they are prodded to do is to establish a calculational relation between measurements and general statements. For example, a climatologist's statement about average temperature may be connected to a set of temperature measurements by the mathematical operation of averaging, and a theorist's formula for adiabatic cooling may be shown by a calculation to explain the observed drop in temperature of a certain updraft.

It was the forecasters and some would-be forecasters who felt the attraction of science-not-art, the desire to make predictions according to specified procedures. From the mid 19th century on, there was great public demand for weather forecasts, yet meteorologists were not content with the fact that forecasting was, as they often put it, "an art rather than a science," and they made repeated attempts to formulate a set of rules for weather forecasting and to base the predictions on the laws of physics. Because forecasting was not perceived as scientific, many meteorologists of the late 19th century abjured the practice, and the British Meteorological Office, for precisely this reason, even stopped issuing forecasts for more than a decade. The efforts to use data in a systematic way and to draw on physical laws for making forecasts often involved much calculation and the use of calculating aids. So algorithms became increasingly important to forecasters too.

There were, of course, other forces leading to an increased use of algorithms—a number of them are discussed in the following chapters—but the strongest ones, and the ones primarily responsible for the new unification of meteorology, were data push, theory pull, and the attraction of science-not-art.

Part I | Meteorology in 1900

As we survey the progress in this department of knowledge, we can discern three collateral aspects: first, the preservation of the memory of the events of past weather and their sequence . . . ; second, speculations upon the relations of those events and upon their proximate and ultimate causes . . . ; and, thirdly, the endeavours to use existing knowledge for the anticipation of future weather. . . . To-day we recognise the corresponding division of labour in modified forms as between the observer . . . , the natural philosopher . . . , and the practical meteorologist. . . .

William Napier Shaw, 1926

At the turn of the century, meteorology encompassed a great variety of studies, but there were three main channels of activity: the empirical tradition of climatology, the theoretical tradition of physics of the atmosphere, and the practical tradition of weather forecasting. As the flow swelled, these channels deepened and diverged.

Chapter 2 An Empirical Tradition

Climatology

Quantitative Description

The modern empirical tradition in meteorology might be traced back as far as William Merle, a fellow of Merton College, who noted the weather at Oxford each day from 1337 to 1344, or to the Danish astronomer Tycho Brahe, who kept daily meteorological records from 1582 to 1598. But it was not until the late 17th century, after the invention of the thermometer and the barometer, that systematic observation became at all common. The Accademia del Cimento of Florence gathered meteorological observations in the 1650s and 1660s. In Paris, soon after the founding of the Académie des Sciences in 1666, regular observations were made at the Academy's observatory. In England John Locke, from 1666 to 1692, made a daily record of temperatures, barometric pressures, and winds, and the Royal Society showed interest in systematic observation. In the 18th century many gentlemen of the Enlightenment kept weather journals, among them George Washington and Thomas Jefferson,[1] and near the end of that century the first network of observing stations, the Societas Meteorologica Palatina, was established by the Elector Karl Theodor of Palatinate-Bavaria.

The increasing interest in meteorological observation was partly the result of a radical change in its nature: from description almost completely qualitative in 1600 to description largely quantitative in 1700. In the ancient and medieval Occident, records of meteorological observations were entirely verbal; rainfall seems not to have been measured, and even wind direction was described categorically rather than numerically.[2] But in the 17th century temperature, atmospheric pressure, humidity, amount of precipitation, wind direction, and wind force all came to be measured. The first thermometer may have been built shortly before 1600 by Galileo, but Santorio Santorre was in 1612 the first to mention such a device in print. Torricelli built the first barometer in 1643. Although hygrometers were devised as early as the 15th century, hardly any were used before the 17th century, and then a great variety of types were built and used. Devices for measuring rainfall, wind direction, and wind force also were constructed in that century. The new instruments gradually brought about a transformation of the science, stimulating a new and greatly expanded observational enterprise and raising new theoretical questions.[3]

Successful quantification of the elements of the weather required more than devices to generate numbers: the numbers had to mean the same, or at least be interconvertible, when generated at different times, at different places, and by different people. In about 1650 Ferdinand II, Grand Duke of Tuscany, made a thermometer whose readings did not depend on atmospheric pressure; and three scientists of the early 18th century—Gabriel Fahrenheit, Réné de Reaumur, and Anders Celsius—showed how to standardize the readings of thermometers. The barometer did not become a reliable scientific instrument until late in the 18th century; Jean André Deluc's calculation of temperature corrections was an important step toward this achievement. Reliable hygrometers first became available in the mid 19th century, thanks in part to methodical studies of hygrometry carried out by Johann Heinrich Lambert and Horace Benedict de Saussure in the preceding century. Most work on the anemometer took place in the 19th century, notably by T. R. Robinson and W. H. Dines; by 1900 both wind speed and direction could be accurately measured.[4]

Reliable instruments were a necessary but hardly a sufficient condition for the communality of data—the possibility for a meteorologist to use with confidence the data gathered by any other meteorologist. This required international agreement about which instruments to use, about calibration of instruments, about procedures for taking readings,[5] and about the recording and communication of data. By the end of the 19th century such agreements had been reached. This was an important aspect of a second transformation of the science, an organizational transformation that occurred mainly in the latter half of the 19th century with the establishment of national weather services, professional societies, and international cooperation among meteorologists. Most national weather services enforced uniformity in the taking and recording of observations, and the principal objective of most 19th-century international meetings of meteorologists was to work toward communality of data.[6]

A look at the forms used for recording meteorological data reveals both the standardization achieved in the late 19th century and the fact that in the past hundred years there has been little change in the sort of observational information gathered. In 1874 an international commission designed standard forms for the recording of meteorological data—some for the actual readings taken and others for presenting summaries—and they were soon in use worldwide. In 1932 R. G. K. Lempfert, president of the Royal Meteorological Society, reported "The old international form of 1874 has stood the test of time well. Naturally, it has undergone changes and development as the years have gone by, but these modifications have generally been of the nature of additions" (p. 95). A similar form, Form 1009, was used in the United States with few changes from 1891, when the Weather Bureau was established as a civilian agency, through 1948.[7] Long-lasting and international agreement about what to observe, along with agreement about how to observe, has contributed greatly to the coherence of the meteorological tradition.

Calculational Demands

The communality of data necessitated, however, a frightful amount of computation. One problem was that in different countries different units of measure were in use. Temperature was measured in degrees Fahrenheit, centigrade, and Reaumur. There were four common barometrical scales: English, Old French, Metric, and Russian. Humidity, speed, weight, length, altitude, and surface area were each measured in a variety of units. A second problem was that corrections often had to be applied to the observed readings. Barometric readings, for example, were regularly corrected for the effects of temperature and capillary action. A third problem was that some quantities were measured indirectly, their values being computed from the observed values of other quantities. Thus, humidity was regularly computed from the observed dew point or from the readings of a wet-bulb and a dry-bulb thermometer, and altitude was often determined by measuring barometric pressures. A fourth problem was that actual values often needed to be converted to corresponding values under standard conditions or at standard times, as converting the actual barometric pressure to the corresponding pressure at 0°C or the actual temperature to the corresponding temperature at sea level.

By 1900 these tasks were being dispatched expeditiously with the help of a computing device—numerical tables—that made each calculation almost as easy as recording the raw data. Although the use of tables as computing devices has a long history in astronomy,[8] it did not become common in meteorology until the second half of the 19th century.

In 1849 Joseph Henry, Secretary of the Smithsonian Institution, persuaded the telegraph companies to transmit weather reports free of charge and began building a network of weather stations. Within a year 150 stations were reporting, and within 10 years 500 stations were reporting. Henry asked Arnold Guyot, Professor of Geology and Physical Geography at the College of New Jersey (renamed Princeton University in 1896), to prepare a collection of tables to be used by the weather observers. In 1852 the first edition of *Tables, Meteorological and Physical* appeared, and subsequent editions appeared in 1857, 1859, and 1884.[9] Guyot wrote in the preface to the first edition:

> The reduction of the observations and the extensive comparisons, without which Meteorology can do but little, require an amount of mechanical labor which renders it impossible for most observers to deduce for themselves the results of their own observations. This difficulty is still further increased by the diversity of the thermometrical and barometrical scales which Meteorologists . . . choose to retain. . . . To relieve the Meteorologist of a great portion of this labor, by means of tables sufficiently extensive to render calculations and even interpolations unnecessary, is to save his time and his forces in favor of science itself, and thus materially contribute to its advancement.

Meteorologists had long used tables to present data. Guyot's tables, on the other hand, are computing aids. The fourth edition, which Guyot had very nearly com-

Reading of Thermometer, Fahr.		Temp of Dew-Point, Fahr.	Force of Vapor in English Inches.	Weight of Vapor.		Humidity, Saturation = 1.000.	Weight in Grains of a Cubic Foot of Air. Height of the Barometer in English Inches.						
Dry.	Wet.			In a Cubic Foot of Air.	Reqd. for Sat'n. of a Cubic Ft. of Air.		in. 28.0	in. 28.5	in. 29.0	in. 29.5	in. 30.0	in. 30.5	in 31.0
°	°	°	in.	gr.	gr.		gr.	gr.	gr.	gr.	gr.	gr.	gr.
62	62	62.0	0.559	6.25	0.00	1.000	491.2	499.9	508.7	517.5	526.3	535.1	543.9
	61	60.3	0.528	5.91	0.34	0.946	491.4	500.1	508.9	517.7	526.5	535.3	544.1
	60	58.6	0.499	5.58	0.67	0.893	491.5	500.2	509.0	517.8	526.6	535.4	544.2
	59	56.9	0.472	5.27	0.98	0.843	491.7	500.4	509.2	518.0	526.8	535.6	544.4
	58	55.2	0.445	4.99	1.26	0.798	491.9	500.6	509.4	518.2	527.0	535.8	544.6
	57	53.5	0.421	4.70	1.55	0.752	492.0	500.7	509.5	518.3	527.1	535.9	544.7
	56	51.8	0.397	4.44	1.81	0.710	492.1	500.7	509.5	518.4	527.3	536.1	544.9
	55	50.1	0.375	4.19	2.06	0.670	492.2	500.9	509.7	518.5	527.4	536.2	545.0
	54	48.4	0.354	3.95	2.30	0.632	492.4	501.1	509.9	518.7	527.6	536.4	545.2
	53	46.7	0.333	3.72	2.53	0.595	492.5	501.3	510.1	518.9	527.7	536.5	545.3
	52	45.0	0.315	3.52	2.73	0.563	492.7	501.5	510.3	519.1	527.9	536.7	545.5
	51	43.3	0.297	3.31	2.94	0.530	492.8	501.6	510.4	519.2	528.0	536.8	545.6
	50	41.6	0.280	3.13	3.12	0.501	492.9	501.7	510.5	519.3	528.1	536.9	545.7
	49	39.9	0.263	2.95	3.30	0.472	493.0	501.8	510.6	519.4	528.2	537.0	545.8
	48	38.2	0.248	2.77	3.48	0.443	493.1	501.9	510.7	519.5	528.3	537.1	545.9
	47	36.5	0.234	2.61	3.64	0.418	493.2	502.0	510.8	519.6	528.4	537.2	546.0

Figure 1 This is part of one page of a 33-page table in the fourth edition of Guyot's *Tables, Meteorological and Physical.* The bottom row tells that the following correspond: (1) dry-bulb temperature 62°F, wet-bulb temperature 47°F, (2) dew point 36.5°F, (3) vapor pressure 0.234 inches of mercury, (4) water-vapor mass 2.61 g per cubic foot, (5) 3.64 g of water vapor required per cubic foot for saturation, (6) relative humidity 0.418, (7) total mass of one cubic foot of air, including the water vapor, 493.2 g for barometric pressure of 28.0 inches, etc.

pleted when he died in 1884, contains about 700 pages of tables (Guyot, 1884). Roughly half of them are for converting units of measure. Of the others, some are for making corrections to the instrumental readings, some for computing quantities measured indirectly, and some for converting the actual values to the corresponding values under standard conditions. Figure 1 shows part of a table for relating various measures of the amount of water vapor in the air; by means of it one could measure humidity indirectly (from wet- and dry-bulb readings or from the dew-point reading) and determine the total mass of a volume of air (knowing the humidity and the barometric pressure).

In 1890 there appeared the very important *International Meteorological Tables Published in Conformity with a Resolution of the Congress of Rome, 1879.* The tables were prepared mainly at the Bureau Central Météorologique in Paris, and English, French, and German editions of the collection were published simultaneously (Air Ministry, England, Meterological Office, 1921, p. 3).

The historian Theodore Feldman has shown that an important motivation in the late 18th century for the calculation and publication of tables for computing elevation on the basis of barometric and thermometric measurements was to allow people without scientific training, primarily the nonmathematical mountaineers, to carry out the calculation (Feldman, 1984, pp. 176–177). Certainly in the latter half of the 19th century the breaking down of procedures into simple steps, such as arithmetical operations and the use of tables, often came from an appreciation of the *horror mathematicae* of the audience. This careful specification of an algorithm is seen in collections of tables such as Guyot's, in observer's handbooks, in training manuals, and in practically oriented textbooks.[10] It is significant that this specification of an algorithm, besides making it easy to use meteorologically untrained people as calculators, drew attention to the individual steps of a calculation for which tables, graphical aids, or other calculating aids might be used, and thus led to greater use of calculating aids.

Trying to Find Regularities by Tabulating

In the course of the 19th century, meteorological data came to be collected by more and more people, with interests as diverse as preventing disease, improving agriculture, benefiting maritime commerce, and contributing to scientific understanding. As a result, the amount of reliable data increased at an accelerating pace, and a few people began to be bothered by the fact that not much use was being made of these data. As early as 1839 the German meteorologist Heinrich Dove wrote, "Lack of material is not so much an obstacle to progress as the inadequate utilization of the data already at hand" (in Landsberg, 1964, p. 137). In 1865 Joseph Henry wrote, "There is, perhaps, no branch of science relative to which so many observations have been made and so many records accumulated, and yet from which so few general principles have been deduced" (in Fleming, 1990, p. 148). Later in the century the astronomer George Airy opined that "the observing is out of all proportion to the thinking in meteorology," and the physicist Arthur Schuster suggested at a meeting of the British Association for the Advancement of Science that meteorologists stop observing for 5 years and work instead to make out what the observations meant (Brunt, 1944, p. 8; Ashford, 1985, p. 47). The piling up of data around them stimulated the meteorologists to devise ways to discover or impose order, which is to say that "data push," the desire to make something out of all the numbers, gave rise to new data-handling procedures.

The most straightforward way was to tabulate the observations—both of the weather and of other phenomena one suspected to be correlated with the weather[11]—and look for a pattern. This was the method advocated by Robert Hooke in 1667 (1667/1958, pp. 172–179). Figure 2 shows a tabular form that he proposed. The rightmost column is headed "General Deductions to be made after the side is fitted with Observations." Hooke thought that merely by looking at the tabulated observations one would see regularities, such as that barometric pressure fell as wind force increased.

It proved difficult, however, to find any regularities. The most determined attempt to make discoveries in this Baconian manner was that of the Palatine Meteorological Society, which hoped to find relationships between celestial motions

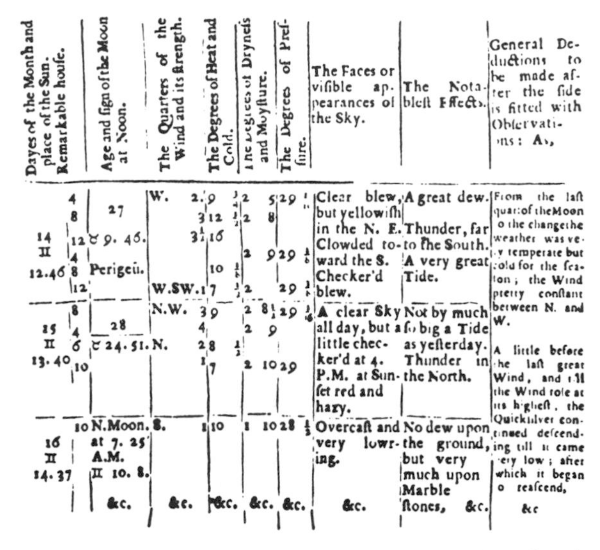

A

SCHEME

At one View repreſenting to the Eye the Obſervations of the Weather for a Month.

Dayes of the Month and place of the Sun. Remarkable houſe.		Age and ſign of the Moon at Noon.	The Quarters of the Wind and its ſtrength.	The Degrees of Heat and Cold.	The Degrees of Dryneſs and Moyſture.	The Degrees of Preſſure.	The Faces or viſible appearances of the Sky.	The Notableſt Effects.	General Deductions to be made after the ſide is fitted with Obſervations: As,
14 ♊ 12.46	4 8 12 4 8 12	27 ♉ 9. 46. Perigeü.	W. 2. 3 3½ W.SW. 1	9 ½ 12 ½ 16 10 ½ 7 ½	2 5 2 8 2 9 2	29 ½ 29 ⅙ 29 ⅙	Clear blew, but yellowiſh in the N. E. Clowded toward the S. Checker'd blew.	A great dew. Thunder, far to the South. A very great Tide.	From the laſt quar: of the Moon to the change the weather was very temperate but cold for the ſeaſon; the Wind pretty conſtant between N. and W.
15 ♊ 13.40	8 4 6 10	28 ♉ 24. 51.	N.W. 3 4 N. 2 1	9 8 ½ 7	2 8½ 2 9 2 10	29 ⅙ 29	A clear Sky all day, but a little checker'd at 4. P.M. at Sun-ſet red and hazy.	Not by much ſo big a Tide as yeſterday. Thunder in the North.	A little before the laſt great Wind, and till the Wind roſe at its higheſt, the Quickſilver continued deſcending till it came very low; after which it began to reaſcend,
16 ♊ 14.37	10	N. Moon. at 7. 25 A.M. ♊ 10. 8.	S. 1	10	1 10	28 ⅓	Overcaſt and very lowring.	No dew upon the ground, but very much upon Marble ſtones,	
		&c.	&c.	&c.	&c.	&c.	&c.	&c.	&c

Figure 2 This is the tabular form Hooke proposed for the recording of weather observations and deductions.

and the weather. From 1780 to 1795 this group of observers from 37 stations in many different countries sent data, taken by standard instruments and recorded on standard forms, to Mannheim in southern Germany where they were tabulated and published *in extenso*. Almost no astronomical correlations of any sort were found.[12]

We may conclude, then, that besides being a satisfactory way to store information, the tabulation of data was forceful in diminishing belief in virtually all simple correlations involving meteorological phenomena. This was, however, an entirely negative role for empirical meteorology. In the 19th century a positive role was found in the establishment of a new, essentially descriptive enterprise that came to be known as climatology. The emergence of that science depended, however, on the development of new ways of treating large amounts of data.

Finding Regularities by Mapping

A quite different approach was to treat the data pictorially: to put the numbers into pictures and then seek visual rather than numerical regularities. The most fruitful form of this strategy was placing data on maps. The first meteorological map was one published by Edmond Halley in 1686. Halley drew a map showing the prevailing maritime winds in and near the tropics, "whereby 'tis possible the thing may be better understood, than by any verbal description whatsoever" (1686, p. 163). This map (Figure 3) clearly reveals some regularities in the prevailing winds.

Although the map was widely noticed and praised, 150 years passed before the drawing of meteorological maps became at all common. There was, for example, not a single map in the 12 volumes of data published by the Palatine Meteorological Society (Cassidy, 1985, p. 22). The dearth of institutional support for meteorology and the quite limited geographic spread of the available data were principal causes of this delay.

In 1817 Alexander von Humboldt introduced a way of picturing the distribution of heat over the earth's surface: on a map of part of the northern hemisphere he drew lines, which he called isotherms, that joined places having the same average temperature. This was the first example in meteorology of a mapping technique that became of great importance to the science. The technique came from the study of terrestrial magnetism and originated from another of Halley's maps, a magnetic chart published in 1701, on which Halley had drawn isogonic lines, that is, lines joining places having the same magnetic declination (the angle between geographic north and magnetic north). In 1721 William Whiston constructed a map with lines of equal magnetic inclination (the angular deviation from horizontal), and in 1804 Humboldt constructed a map with lines of equal magnetic intensity (Hellman, 1895). It was thus from the study of terrestrial magnetism that Humboldt imported the technique into meteorology.

Humboldt's essay presenting the isothermal map was read to the Académie des Sciences in May and June 1817, and within 2 years extracts appeared in four journals in three languages. It had a great impact on meteorology, as the isoline method came to be applied to many different phenomena (Robinson and Wallis, 1967, p. 122; Landsberg, 1964, p. 131). Heinrich Dove constructed maps with monthly isotherms (the lines joining places having the same mean temperature for a given month), introduced the concept of isothermal surfaces, and constructed maps with other types of isolines. By 1870 other meteorologists, notably H. W. Brandes and L. F. Kaemtz, were drawing isolines for barometric pressure, deviation of pressure from normal, annual precipitation, and frequency of thunderstorms, as well as for temperature and deviation of temperature from normal.

Maps of barometric pressure proved especially useful. Kaemtz, for example, by charting barometric variability showed clearly that the north Atlantic is characterized by barometric instability, equatorial regions by constancy (Landsberg, 1964, p. 132). Of paramount significance was the discovery that in the north temperate zone low-pressure regions, which are often regions of precipitation, move fairly regularly from west to east. In fact this predictability of the motion of storms, together with the means, provided by the telegraph, to construct same-day weather maps, was the most important factor in the establishment of national weather ser-

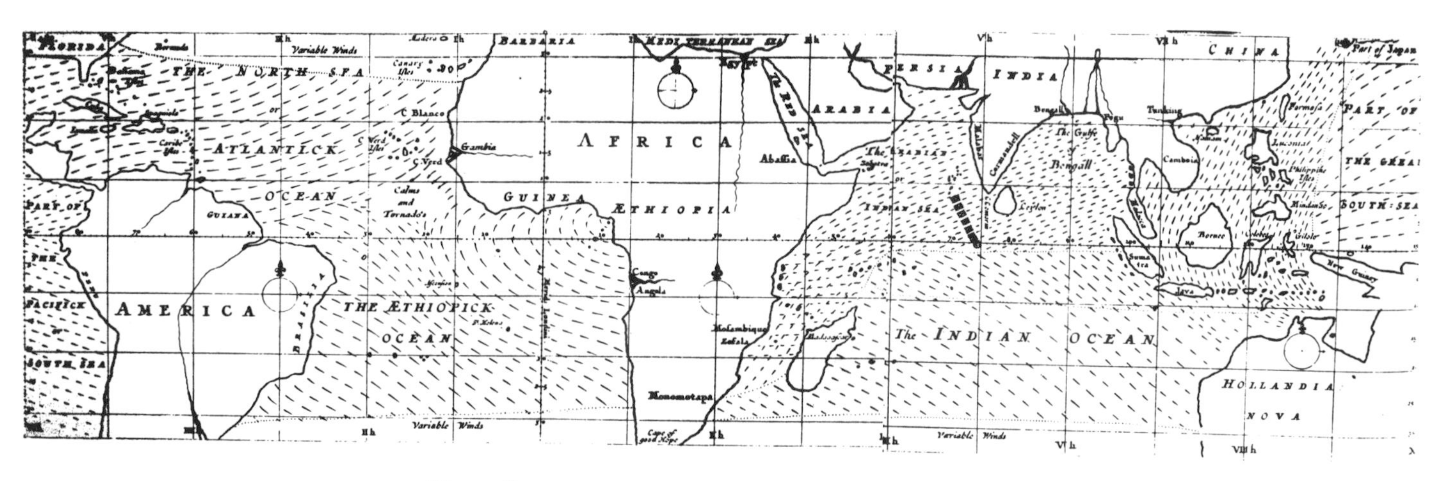

Figure 3 Halley's map (1686) of the prevailing maritime winds.

vices in the 1860s and 1870s. The history of such weather maps is discussed more fully in Chapter 4.

Another type of map also had extraordinary practical utility. In the 1840s the American naval officer Matthew Fontaine Maury collected data from ships' logs and began publishing charts of winds and currents. The use of Maury's charts in navigation markedly shortened sailing times. The average passage from New York to San Francisco was reportedly reduced from $187\frac{1}{2}$ days to $144\frac{1}{2}$ days by the use of Maury's charts, and comparable reductions were claimed for routes worldwide (Williams, 1963, p. 192). Maury soon became an international celebrity. His work attracted the attention of national leaders wanting to promote maritime commerce and abetted government involvement in meteorology. For example, Maury gained the official support of 10 or so countries in arranging for the first international conference of meteorologists, which took place in Brussels in 1853.

At about the same time James Coffin, availing himself of observations gathered by the Smithsonian Institution as well as Maury's data, constructed a series of wind charts that were published in 1853 (in "On the winds of the Northern Hemisphere") and in 1875 (in "The winds of the globe: or the laws of atmospheric circulation over the surface of the earth"). Coffin's charts formed the empirical basis for some of the important work of the theoretician William Ferrel, which is considered in Chapter 3 (Fleming, 1990, p. 136).

Because maps proved to be an effective way to present large amounts of data in a concise form, climatologists, moved by "data push," made them their principal means of interpreting meteorological information. Climatological atlases became common; one of the first was Heinrich Berghaus's *Physikalischer Atlas* published in 1845. Maps came to be extensively used in textbooks of climatology; W. G. Kendrew's widely used *The Climates of the Continents* (1922) includes about a hundred maps. In 1944 David Brunt said, "The distribution of surface observations of temperature, pressure and wind over the globe has been investigated largely by means of charts [that is, maps] of monthly mean values of these factors . . ." (p. 5).

Finding Regularities by Graphing

Another way of dealing with "data push" was provided by graphs. It is remarkable that a method of data interpretation as prominent as graphing today is was hardly used a hundred years ago. Of course the correspondence between algebraic equations and curves goes back at least to Descartes, and data were occasionally plotted on coordinate systems as early as the second half of the 18th century (Frisinger, 1977, p. 83; Cassidy, 1985, p. 21).[13] Adolphe Quetelet published graphs of meteorological data in 1827, as did Francis Galton in 1863. But the method did not become common until near the end of the century.

Neither the noun "graph," in the sense of a mathematical curve, nor the verb "to graph," in the sense of producing such a curve, was used before the 1880s.[14] Amid all the tables and maps there are virtually no graphs in the *Monthly Weather Review,* which began publication in 1873, until the mid 1880s, when graphs begin to appear in considerable numbers. Julius Hann's 1883 *Handbuch der Klimatologie* contains only a few graphs, and even these are part map in that one coordinate axis represents latitude.[15] As graphing became common among scientists in the 1890s, new types of graphs appeared; for example, it was soon a standard technique to plot data on logarithmic graph paper, that is, graph paper with logarithmically spaced rulings in one or both directions, and such paper became commercially available (Boys, 1895, p. 272). Pantographs were manufactured for the copying of graphs, and more complicated mechanical devices were built for transforming graphs (Shaw, 1934, p. 104). Like numerical tables, graphs served as records of observations, as aids in interpreting the data, and as calculating aids.

One of the most common graphs in the meteorological literature early in this century was a graph of temperature as a function of elevation. In 1817 Humboldt presented, along with his map of isotherms, a diagram showing the decrease of temperature with height, and this diagram was used to calculate the sea-level temperature corresponding to the temperature of a station not at sea level. Thus Humboldt's essay is historically important for presenting one of the very first meteorological maps and one of the very first graphical procedures for meteorological calculation.[16] In the first years of the 20th century, graphs of temperature as a function of elevation made clear the existence of a different temperature regime—later named the stratosphere—above a height of about 10 km (see, e.g., the graph on p. 264 of Shaw, 1926).

Climatologists were quite inventive of other sorts of visual presentation of data. Because in many parts of the world there is a high correlation between wind direction and weather type, the graphing of the frequencies of wind in different directions was often useful. So-called wind roses, besides showing the directional distribution of winds, showed weather elements, such as humidity or cloudiness, associated with each wind direction. A more sophisticated example, taken from Willis Moore's *Descriptive Meteorology* (1914), is shown in Figure 4. It indicates the prevailing wind direction at Chicago for each hour of the day and for each month of the year. Superimposed on this grid are isolines of the departure of temperature from the daily mean; the thick lines show the time of day when the actual temperature equals the daily mean (once in the late morning and once in the evening). The dotted lines show the time of sunrise and sunset.

Quite a few climatic features of Chicago can be read from this figure. For example, the arrows show that in the summer the morning southwest wind is followed by an afternoon northeast wind, which is a sea breeze off Lake Michigan, while in the late fall and the winter the prevailing westerlies are little affected by the presence of the lake. The isolines show that it is coolest just before dawn and warmest a few hours before sunset, and that the temperature changes most rapidly in the hours before noon.

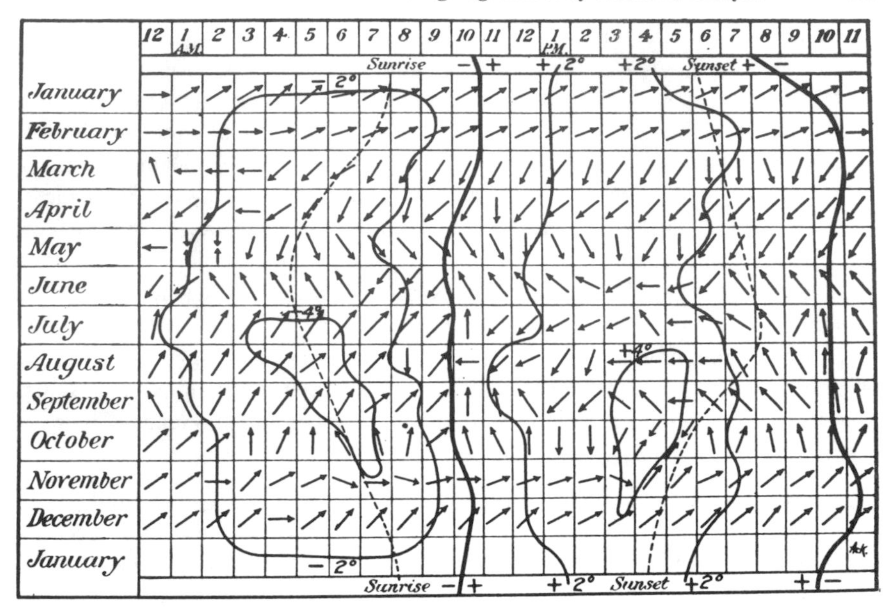

Figure 4 The horizontal axis represents hour of the day, the vertical axis month of the year. Arrows show prevailing wind direction, solid lines are isolines of departure of temperature from the daily mean, and dotted lines indicate times of sunrise and sunset (Moore, 1914, p. 183.)

Finding Regularities by Statistical Analysis

In the early 19th century many people became interested in the systematic gathering of information about a country and its inhabitants. This interest lies behind the etymology of "statistics," which meant originally the political studies dealing with numerical facts about a state. One type of information sought was meteorological data. In Prussia the first meteorological institute was established as a branch of the state statistical office; in England in the 1840s the statistical office of the Registrar General asked James Glaisher at the Greenwich Royal Observatory to submit meteorological data for its study of the relation between the weather and the death rate and other social indicators; and the Italian statistical bureau gathered meteorological data (Khrgian, 1959, p. 111; Sheynin, 1984, p. 76). Practical interests, such as in commerce, agriculture, forestry, and public health, and scientific interests, such as in astronomy, magnetism, and botany, motivated the gathering of the meteorological data. The institutions carrying it out were numerous and included state statistical offices, state agricultural offices, military agencies, astronomical observatories, and various scientific organizations. As state meteorological services became established and grew—almost every European country had a meteorological service by 1875—they assumed more and more of the task of data collection, and indeed in the latter part of the 19th century the main activity of many weather bureaus was simply the compilation of data.

This data gathering gave a continuing push to the new science of climatology, and climatology became in large measure a statistical science. Indeed, the very concept of climate came to be defined in statistical terms. Julius Hann wrote, "By climate we understand the totality of meteorological phenomena which characterize the average state of the atmosphere at some place on the earth's surface" (p. 1).[17] As this quotation suggests, a main concern of climatologists was simply the calculation of daily, monthly, and yearly averages of the elements of the weather.

Climate came, however, to be understood in a broader sense as including the extremes of weather and the variability of the weather, on various times scales. Consider Julius Hann's presentation in 1883 of the climate of Vienna (Figure 5). First, the information is entirely quantitative. According to Hann,[18]

> A scientific climatology must strive to express all climatic elements in numerical values, because it is only through actual measurements that we can achieve directly comparable expressions and exact conceptions of meteorological circumstances and states. (p. 4)

Second, almost all the numbers are mean values; indeed, the only exceptions are the extreme temperatures listed near the right in the first table. Third, although most of the numbers express the (monthly or yearly) average weather, some express the variability. The third column of the first table, for example, gives the mean deviation from the mean temperature.

The broadened conception of climate was expressed in 1911 by A. Hettner as follows:

> But with all its changeability the weather at each place on the earth's surface has a definite general character, which expresses itself not only in average and extreme values, but also in the nature of the change, in the entire periodical and non-periodical course of the weather. This totality of weather phenomena of a place we call its climate (in Schneider-Carius, 1955, p. 334).[19]

As this quotation suggests, it became common to look for and find periodicities—diurnal, annual, and of many other lengths—and to separate periodical from non-periodical change. The search for weather cycles reached its greatest intensity in the first two decades of the 20th century, and then, in the 1920s and 1930s, almost came to an end. The story of this tradition, and of the graphical, numerical, and mechanical techniques of calculation that were therein employed, is deferred until Chapter 8.

With climate so conceived, it is not then surprising that climatologists made use of statistical techniques. For the most part, however, they confined themselves to the computation of averages. Some of them did discuss the distribution of a set of data—as being, for example, normally distributed with known mean and standard deviation—and some of them did develop new statistical techniques—in 1888

TABLE I.

CLIMATIC DATA FOR VIENNA, LAT. 48° 12′ N., LONG. 16° 22′ E., ALTITUDE ABOVE SEA LEVEL, 194 METERS.

A.—TEMPERATURE (in degrees Centigrade).

	Mean derived from 24-hour observations (20 years).	100-year Mean (reduced)	Mean departures of the Means (90 yrs.).	Means for three observation hours (20 years).			Diurnal range of temperature (20 years).		Mean monthly and annual Extremes (20 years).		Mean monthly and annual Range.	Absolute Extremes, 1829-1875.		Mean diurnal variability of temperature.
				6 a.m.	2 p.m.	10 p.m.	Periodic.	Non-periodic.						
	°	°	°	°	°	°	°	°	°	°	°	°	°	°
December, . .	−0·8	−0·3	2·3	−1·5	0·6	−1·0	2·1	4·7	9·6	−11·2	20·8	19·1	−22·6	2·0
January, . .	−1·3	−1·7	2·5	−2·3	0·3	−1·6	2·7	4·9	9·7	−12·1	21·8	18·8	−25·5	2·1
February, . .	0·4	0·1	2·2	−1·2	2·6	0·1	3·8	6·1	11·4	−10·0	21·4	20·0	−20·0	2·0
March, . . .	4·2	4·3	1·8	1·6	7·4	3·6	5·9	7·8	16·7	−5·9	22·6	24·3	−13·3	1·8
April, . . .	10·0	9·9	1·7	6·2	14·0	9·0	7·8	9·6	23·9	−1·0	24·9	28·8	−7·0	1·9
May, . . .	15·1	15·1	1·5	11·4	19·3	13·8	8·2	10·2	28·5	2·7	25·8	36·0	−1·6	1·8
June, . . .	18·6	18·8	1·2	15·5	22·4	17·1	7·6	9·9	31·5	9·1	22·4	37·8	3·8	1·9
July, . . .	20·3	20·5	1·3	16·9	24·3	18·9	7·9	10·1	32·6	11·0	21·6	38·8	8·0	1·9
August, . .	19·6	19·7	1·3	16·0	23·7	18·2	7·9	9·7	32·9	9·8	23·1	37·5	5·6	1·8
September, . .	16·1	15·9	1·2	12·2	20·4	14·8	8·2	9·6	28·3	4·9	23·4	33·5	−0·6	1·7
October, . .	10·5	10·0	1·4	7·7	14·3	9·5	6·6	8·3	23·2	0·6	22·6	27·1	−6·8	1·5
November, . .	3·7	3·9	1·4	2·5	5·5	3·3	3·1	4·9	14·9	−5·9	20·8	21·3	−15·0	1·8
Year, . . .	9·7	9·7	0·74	7·1	12·9	8·8	5·9	8·0	33·9	−15·1	49·0	38·8 July 14, 1832	−25·5 Jan. 22, 1850	1·9
Column, . .	1a	1b	1c	2	3	4	5	6	7	8	9	10	11	12

TABLE II.

CLIMATIC DATA FOR VIENNA—CONTINUED.

B. HUMIDITY, PRECIPITATION, CLOUDINESS, WIND VELOCITY, EVAPORATION, OZONE.

	Humidity (20 years).					Rain and Snow.		Cloudiness 0 - 10 (20 years).	Mean Duration of Sunshine in Hours.	Ditto in Percentages of greatest possible Duration.	Mean Wind Velocity (meters per second).		Evaporation (5 years) (mm.)	Ozone 0 - 10 (20 years).	
	Mean Vapour Tension. (mm.)	Relative Humidity (in percentages).			True Mean.	Amount (34 years). (mm.)	No. of Rainy Days (20 years).		(15 years).		City (6 years).	Observatory (20 years).		Day.	Night.
		6 a.m.	2 p.m.	10 p.m.											
December, ·	3·7	86	77	86	83	40	12·4	7·3	49·1	19	2·4	5·2	18	3·1	5·5
January, ·	3·6	87	77	86	84	35	12·8	7·2	65·5	23	1·7	5·1	13	3·2	5·8
February, ·	3·8	84	70	83	80	36	11·2	6·7	86·5	30	2·6	5·4	27	4·2	6·0
March, · ·	4·4	81	58	76	71	43	13·1	6·2	129·8	36	2·2	6·2	39	4·2	6·2
April, · ·	5·7	76	48	68	63	42	12·3	5·2	178·9	43	2·4	5·2	71	4·6	5·7
May, · ·	8·2	76	49	71	64	64	13·0	5·1	240·9	51	2·0	5·4	87	5·2	5·4
June, · ·	10·0	75	50	71	64	66	12·7	4·9	228·2	48	2·4	5·3	93	5·2	5·6
July, · ·	10·9	75	48	70	63	65	13·3	4·5	274·0	56	2·2	5·5	113	5·3	5·3
August, ·	11·0	79	50	73	66	72	11·8	4·5	246·1	56	2·1	4·9	94	5·1	5·4
September, ·	9·3	82	53	75	69	45	8·3	4·5	175·9	48	2·0	4·7	77	3·9	4·5
October, ·	7·2	85	61	81	76	44	10·6	5·4	100·5	30	2·0	4·6	47	3·0	4·3
November, ·	4·8	84	72	83	80	43	12·6	7·4	63·2	23	3·0	4·9	32	3·0	5·1
Year, · ·	6·9	80	59	77	72	595	144·1	5·7	1838·6	41·4	2·2	5·2	711	4·2	5·4
Column, ·	1	2	3	4	5	6	7	8	8 b	8 c	9	9 b	10	11	12

Figure 5 Two tables, published in the English translation of Julius Hann's *Handbuch der Klimatologie* (1883), that present climatic data for Vienna.

Wladimir Köppen introduced a measure of the asymmetry of a frequency distribution (Schneider-Carius, 1955, p. 328). Indeed several of the people who figure prominently in the history of statistics, such as Adolphe Quetelet, Francis Galton, and Arthur Schuster, did much work with meteorological data.

Of the new techniques, the method of correlation was perhaps most important. Introduced in about 1890 by the English gentleman-of-science Francis Galton,[20] the method allows one to compute a number that indicates how much of the variation in one quantity can be accounted for by the variation in another quantity. Although the method, which came to be used by many meteorologists, was entirely arithmetical, Galton and others often drew diagrams, where pairs of numbers were plotted on a Cartesian coordinate system, that made clear the import of the correlation coefficients and undoubtedly increased the appeal of the method. Galton devised other graphical procedures, and some of them are to be found in his *Meteorographica, or Methods of Mapping the Weather* (1863).

There was, or course, considerable computational labor involved in processing climatological data. Both the British Rainfall Society, an organization of mainly amateur weather-watchers, and the Meteorological Office had a computing staff for producing the statistical summaries. An indication of the amount of routine computation is provided by Ernest Gold's comment about the adoption in 1929 of the millibar as the sole barometric unit in international data exchange: "The economic advantage through the elimination of the conversion of thousands of values in hundreds of meteorological offices daily is substantial" (Gold, 1945, p. 214). It is interesting that Gold added, "The elimination of the consequent source of mistakes is even more important . . ."; we will see in later chapters that an important motivation for automating data processing was the desire to eliminate human error.

Historical records occasionally include an estimate of the time spent in computation. James H. Coffin, a professor at Lafayette College in Easton, Pennsylvania, contracted in 1855 to prepare daily, monthly, and yearly averages of measurements taken in a special program, jointly administered by the U.S. Patent Office and the Smithsonian Institution, intended to promote agriculture. In the year 1856 Coffin processed some half-million observations. To do this he employed 12 to 15 people on a part-time basis—the equivalent of 3 or 4 people full-time—to reduce the observations to sea level and to 32°F and to compute the averages. Joseph Henry, Secretary of the Smithsonian Institution, estimated that, on average, one minute of calculation was required to process each observation (Fleming, 1988, pp. 289–290).

Although the statistical approach became characteristic of the empirical tradition in meteorology, theoreticians and forecasters also made use of statistical techniques; some of these are described in the following two chapters.

The Establishment of Climatology

In the second half of the 19th century climatology became established as a science.[21] There appeared comprehensive surveys of national climate, such as Adolphe Quetelet's *Climatology of Belgium* (1845–1853), Lorin Blodget's *Climatology of the United States* (1857), and K. S. Veselovskii's *On the Climate of Russia* (1857). There appeared detailed studies of climate worldwide, such as Julius Hann's *Handbook of Climatology* (1883), and A. I. Voeikov's *Climates of the World* (1884). Climatology began to be taught in colleges and universities; thus, in 1903 Robert De Courcy Ward published his translation of Volume 1 of Hann's *Handbook of Climatology* to be used as a textbook for a course taught at Harvard University entitled "General Climatology." And, as we have seen, the study of the meteorological data, when mapped, graphed, or averaged appropriately, revealed regularities in the weather.

As the science was cultivated, even higher levels of generalization proved possible. For example, in Hann's book there are statements such as "The diurnal range of temperature increases with increasing distance from the ocean, as does the annual range" and "In winter, there is everywhere a tendency to the formation of a barometric minimum over the enclosed portions of the oceans, and the occurrence of southerly and westerly winds on the eastern sides, and of northerly off-shore winds on the western sides of large bodies of water."[22] Hann also presented a number of empirically derived formulas giving, for example, mean temperature or temperature difference between warmest and coldest months as a function of latitude (see 1883, pp. 93, 134).

In a 1924 book entitled *Climatic Laws,* Stephen Visher, an American professor of geography, presented 25 meteorological laws and 90 climatological laws. Many of the meteorological laws came from physics,[23] but most of the climatological laws came directly from a study of the data. Virtually all elements of the weather were found to exhibit regularities. For example, the annual range in temperature was found to increase with latitude up to the region of persistent snow; and the frequency of precipitation was found to increase directly with the annual amount but inversely with the monthly range in precipitation. Other of Visher's laws are the following: "Wind velocities average greater in winter than in summer in mid-latitudes," "Precipitation increases with altitude to moderate heights and then decreases steadily until at the height of 2 or 3 miles (3–5 km.) it is slight," and "Many mountain valleys and leeward slopes have a peculiar climate because of local hot winds. . . ."

Finally, the new science led to classifications of climate. The climatic classification that was most used was that of Wladimir Köppen, first presented in 1884. Until 1931 Köppen continually revised his scheme of classification. In his 1918 version he defined the climatic types quantitatively; for example, the border between a steppe climate and a neighboring moist climate was defined to be the locus of points for which $r = 0.44\ t - 8.5$, where r is the rainfall in inches and t is the temperature in degrees Fahrenheit (Khrgian, 1959, p. 329; Petterssen, 1958, p. 291).

All such empirically discovered regularities became standing challenges to the ingenuity of theorists; in Chapter 3 we will see some examples of successful theorizing of this sort. More importantly, however, such regularities soon became valuable explanatory resources for scientists—botanists, zoologists, geologists, sociologists, and others—and proved of practical value, especially for agriculture and commerce.[24] Thus despite the disappointment of finding very few numerical regularities, by the time of World War I the accumulation and processing of meteorological data had become established as an important activity. Already strong in itself, this empirical tradition was further strengthened by its connections with theoretical meteorology and weather forecasting, which are the subjects of the following two chapters.

Chapter 3 A Theoretical Tradition

Physics of the Atmosphere

Aristotle's *Meteorologica,* written about 340 B.C., established a theoretical tradition in meteorology and ensured that, throughout the Western world, meteorology would be studied as a part of natural philosophy. This treatise remained the basis of Western theoretical treatments of meteorology until the early 17th century (Frisinger, 1977, p. 22). In that century, however, theoretical meteorology underwent great changes: Descartes stimulated new thinking about atmospheric phenomena, especially by the publication of *Les météores,* appended to his *Discours de la méthode* (1637); new observations, many of them resulting from the invention of the barometer and thermometer, called for new explanations; and the development of the science of mechanics prompted new theories of atmospheric phenomena, such as Edmond Halley's 1686 theory of the trade winds.

The Beginnings of Dynamical Meteorology

Despite the data provided by thermometer and barometer, and despite the existence of relevant mathematical theory in Newton's *Principia,* until the mid 19th century meteorological theories remained almost entirely qualitative. There were, it is true, a few topics that were treated mathematically, such as the relationship between elevation and barometric pressure or the nature of atmospheric tides. There was a branch of climatology that studied what Hann called "solar or mathematical climate" (1883, pp. 57–79) producing such mathematical results as a formula expressing insolation as a function of geographic latitude. There were also laboratory studies that revealed quantitative relationships between measured variables, such as James Espy's investigations of water vapor in the years around 1850. Most important of the laboratory studies were those aimed at understanding meteorological instruments, as Jean André Deluc's investigations of thermometers in the 1760s, Horace Benedict de Saussure's 1783 *Essais sur l'hygrométrie,* and William Henry Dines's studies in anemometry in the 1880s. On the whole, however, meteorological theory remained nonmathematical until the second half of the 19th century.

In 1851 the French physicist J. B. L. Foucault used an extremely long pendulum to demonstrate the deflecting effect of the earth's rotation. News of Foucault's work stimulated William Ferrel, an American schoolteacher, to analyze the effect of the earth's rotation on atmospheric motions (Kutzbach, 1979, p. 37), and in 1860 Ferrel published in New York and London a small collection of his papers that had appeared separately in the late 1850s. In these papers Ferrel presented a general mathematical theory of fluid motion on a rotating earth and applied this theory to understanding winds and currents (Burstyn, 1971, p. 591). For example, he gave a theoretical explanation of the observation by the Dutch meteorologist C. H. D. Buys Ballot that wind direction is generally parallel to the local isobars and wind speed is roughly proportional to the barometric gradient (that is, proportional to the rate of change of barometric pressure in the direction of fastest change) (Burstyn, 1984, p. 341).[1] Ferrel's work, which was widely noticed, may be said to have initiated what became known as dynamical meteorology, the theoretical treatment of atmospheric motions on the basis of the laws of physics.

In the half century following Ferrel's 1860 publication, dynamical meteorology was cultivated by, among others, Max Margules and Felix Exner in Austria; Hermann Helmholtz, A. Oberbeck, Wilhelm von Bezold, and Adolf Sprung in Germany; Henrik Mohn and C.M. Guldberg in Norway; William and James Thomson in England; and William Ferrel and Cleveland Abbe in the United States. A year or two before World War I Felix Exner set himself the task of summarizing and systematizing the work of these theoreticians, and in 1917 he published *Dynamische Meteorologie,* which may be taken as marking the general acceptance of dynamical meteorology as a discipline.[2]

The attitude of most of the above-named people was that meteorology ought to be applied physics and that observational data ought to be explained deductively. They saw themselves as doing a new kind of meteorology, on a different tack from the empiricists such as Hugo Hildebrandsson in Sweden, Heinrich Dove in Prussia, C. H. D. Buys Ballot in The Netherlands, James Glaisher in England, and Elias Loomis in the United States, who held that meteorology was an independent science whose laws were to be derived inductively from the data. In 1890 Cleveland Abbe complained, "Hitherto, the professional meteorologist has too frequently been only an observer, a statistician, an empiricist—rather than a mechanician, mathematician and physicist,"[3] and in the same year Wilhelm von Bezold wrote that meteorology was being transformed into "physics of the atmosphere" (Kutzbach, 1979, p. 46).

The two attitudes, however, did not result in separate research communities. The applied physicists were not opposed to discovering regularities inductively, nor did the empiricists object to applying physical laws to the atmosphere. For many meteorologists it was simply a question of expediency: is it more fruitful to work from physics or from the data? One might, of course, do both, and many did. Julius Hann, although a vocal proponent of the empirical approach, was one of the first to use thermodynamics in an explanation of atmospheric phenomena. Adolf Sprung, most of whose work proceeds deductively from mathematical physics to the data, also worked in the other direction, inductively. But there were meteorologists who doubted the adequacy of physics, and for them it was not simply a question of expediency. It may be, they thought, that there are meteorological laws not deducible from physics, or not, at any rate, from the known laws of physics. An influential exponent of this view was Napier Shaw.

William Napier Shaw

Shaw, who was born in 1854, studied mathematics, physics, and chemistry for 4 years at Emmanuel College, Cambridge. He then worked for a short period under Clerk Maxwell at the Cavendish Laboratory and under Helmholtz in Berlin. In the 1880s and 1890s he was active in teaching and administration at Emmanuel College and did theoretical and experimental work on hygrometry, theory of ventilation, and other topics. From 1900 to 1920 Shaw directed the Meteorological Office,[4] where he worked hard to bring scientifically trained men and women into the Office for the purpose of conducting research and making forecasting more scientific. David Brunt wrote that Shaw instilled enthusiasm into the Office, lifting it from "the Slough of Despond" it found itself in at the end of the 19th century. Brunt (1951) wrote also, "It is no great exaggeration to say that Shaw found meteorology an exercise in arithmetic, and left it a branch of physics, the contributions which he and his early scientific colleagues in the Office made to the subject being of prime importance" (p. 119).[5]

Shaw was indeed well trained and talented in mathematical physics. He often worked in a purely deductive way, deriving equations of meteorological import from physical laws or investigating the properties of a mathematical model while setting aside the question of the model's fidelity to the physical world.[6] He devised two important forecasting tools—the tephigram and isentropic analysis (both discussed in later chapters)—entirely on the basis of physical principles.

Nevertheless, on many occasions he argued emphatically for taking more observations and for working inductively from the data. He wrote that throughout its history "meteorological theory has been invariably hampered by want of facts" and that this was still true in 1926 even though "the volumes containing the facts about the atmosphere are so numerous as to be quite overwhelming" (Shaw, 1926, p. 316). He thought that it was premature to attempt to calculate the motion of the atmosphere on the basis of the known laws of physics, and that the proper course was first to discover a representation of the actual motions of the atmosphere and then from this infer the underlying laws (Shaw, 1903b, pp. 418–419). Thus he thought that meteorology needed, above all, a Kepler, someone to discern a pattern in the data: "The first great step in the development of any physical science is to substitute for the indescribably complex reality of nature an ideal system that is an effective equivalent for the purpose of theoretical computation" (Shaw, 1903b, p. 418). And he was outspoken in his belief that meteorologists should not assume that the laws of physics suffice:

> It seems more than possible that the true theory of meteorology will never be evolved by the iteration of marginal notes until they fill the page, and that the true course of progress is to accept Maxwell's hint to develop the representation of the

> motion to such a degree of perfection that the forces will be deduced from it, instead of supposing that we can specify the forces and that nothing but the method of fluxions is necessary to deduce the motion. (Shaw, 1926, p. 322)

As this was the final paragraph of his history of meteorology, we may be sure that Shaw felt strongly about the view stated.[7]

The Theorists and the Empiricists

Shaw's attitude toward the data was shared by quite a few meteorologists. J. M. Pernter of the Austrian weather service wrote in 1903 that it was only by knowledge of the actual weather conditions "for every place and for every type of pressure distribution . . . that we can hope at some time to discover the fundamental laws of the changes in the weather" (p. 160). Shared too was Shaw's expectation of the Messiah's coming in the character of a Kepler. Willis Moore, Chief of the U.S. Weather Bureau, wrote in 1898

> When our extensive system of daily observation has been continued for another generation a Kepler or a Newton may discover such fundamental principles underlying weather changes as will make it possible to foretell the character of coming seasons. If this discovery be ever made it will doubtless be accomplished as the result of a comprehensive study of meteorological data of long periods covering some great area like the United States. At any rate we are certainly now laying the foundation of a great system which will adorn the civilization of future centuries. (p. 14)

While for the applied physicists the data served mainly to guide and check theoretical deduction, for the empiricists, as this quotation suggests, the data comprised the substance of meteorology.[8] And while the former thought the data too complex to reward an inductive approach, the latter expected the data to yield a true science just as Tycho Brahe's astronomical data had done.

For most of the 19th century the empiricists held the field. Although quantitative regularity was hard to find in the data, there was enough qualitative regularity to support theorizing. The theories of storms by Heinrich Dove, Robert Fitzroy, William Redfield, James Espy, and many others belong in this tradition. Around the turn of the century, however, the mathematical theories of the dynamical meteorologists became the dominant style. One reason is apparent: one gets an impressively rigorous and quantitative theory if one starts with mathematical physics. More important is that they succeeded in connecting such theories to the data.

Without this connection between theory and data, dynamical meteorology would have been a branch of physics of little interest to the empirical meteorologists. This was never the case, mainly because the dynamical meteorologists saw to it that their work applied to the actual atmosphere. They stressed "contact with reality" and sometimes limited their theorizing to the explanation of observed phenomena.[9] Moreover, some of the practitioners of dynamical meteorology, like

Hann and Loomis and Shaw, were wholeheartedly empiricist by conviction, so did not rest until theory was tested by observation.

One common practice was to start with a physical theory and make it apply to the conditions prevalent in the earth's atmosphere. This was the usual manner of presentation in textbooks of dynamical meteorology. This was the way in which Helmholtz, C. T. R. Wilson, and others used theories of the behavior of air saturated with water vapor to explain the formation of clouds. Often, however, as was the case with these theories of cloud formation, the unavailability of the relevant observational data made the work seem more an exercise in deductive reasoning than an explanation of the actual atmosphere.

Another common practice was to start with an observed phenomenon and then show how the laws of physics might account for it. Thus George Hadley in 1735 used the principle of the conservation of angular momentum to explain the general direction of the trade winds, and Julius Hann in 1866 used thermodynamics to explain the Föhn, a warm dry wind that blows down from the Alps, and Helmholtz in 1888 used fluid dynamics to account for the form of altocumulus clouds. However, the multiplicity of atmospheric processes often made it difficult to be sure that a plausible mechanism for an effect was in fact the main cause.

What gave meteorologists confidence that theory and observation were actually connected—whether the meteorologists started with theory and made it apply to atmospheric conditions, or started with an observed regularity and explained it on the basis of physics—was quantitative agreement, and quantitative agreement came about only through calculation.

Calculation in Theoretical Meteorology

This reaching down to the data to connect theory and measurement—what I have called "theory pull"—made therefore calculation important. There was considerable variety in the way calculation was employed by the theorists, and we will here briefly survey this range. In doing so, we should bear in mind the equivocality of the word "calculation," already noted in Chapter 1. Sometimes it means simply the carrying out of a preestablished algorithm and sometimes it means devising an algorithm as well as carrying it out. (When an engineer "calculates" the wind resistance of a particular structure, it is the latter he is doing.) Since the devising of the algorithm may be more or less predetermined or straightforward, the two meanings grade into each other. The important point here is that when the algorithm is not given in advance, then the devising must be constrained by an accepted theory; we would otherwise speak of an estimate or a judgment rather than a calculation. Indeed, the devising of an algorithm often proceeds by formal deduction, as in a mathematical proof, eked out by explicit *ad hoc* assumptions, such as "the process is effectively adiabatic."

The principal role of calculation has been to provide support for a theory. For example, in 1865 William Thomson explained, by means of a calculation based on thermodynamics, observed values of the decrease of air temperature with height (Kutzbach, 1979, p. 47). Often, as with the "Zahlenbeispiele" (numerical examples) of Margules and Exner, a calculation was carried out only to show the reasonableness of the theory, order-of-magnitude agreement being taken as satisfactory. From the 1870s on, calculations of energy transformations have been common in dynamical meteorology; a theory of storms, for example, was expected to account for the amount of kinetic energy involved (Kutzbach, 1979, p. 90).

Calculations have been more decisive, however, in refuting theories. Jean d'Alembert proposed that the gravitational effects of the sun and the moon caused the trade winds; Laplace showed by a calculation that the effects were too small to be the cause (Gillespie, 1978, p. 301). In 1901 Margules used calculation to refute the theory that the kinetic energy of storms comes from the potential energy of the pressure field.[10]

In these cases a theoretical model, which was well understood, was tested against observational data by means of a calculation. Sometimes, though, the implications of a set of assumptions were unclear and the purpose of calculation was to investigate rather than to test the model. In recent decades this type of calculation has come to be called numerical experimentation, but the practice has a long history.

The essence of laboratory experimentation is that a physical system is placed in a known state and then observed as it changes state. In numerical experimentation a mathematical system is specified at the outset, and the implications of this specification are then worked out. One looks to the physical world only at the outset, as a guide to the specification of the mathematical system, and at the end, to see if the action of the system resembles the action of the physical world.

This is exactly what Napier Shaw did in a 1903 article on the paths of air in a traveling storm. The assumptions are stated at the outset:

> The special case I propose to deal with is that in which the speed of the air is uniform over the area of the storm, although the direction varies from point to point. I shall also suppose the isobars to be true circles and the wind directions tangential to the isobars. Lastly the center will be regarded as describing a straight path with the same speed as the wind at any point.

Shaw then proceeds as a mathematician, remarking that "Whether this ideal state of things represents a possible reality is a matter for subsequent consideration." After working out some of the implications of the model, Shaw does compare it with the physical world: "The trajectories constructed from the recorded directions and velocities of the wind bear such a relation to the path of the center that the applicability of the kinematical reasoning here employed is quite unmistakeable." (1903a, pp. 318, 319)

Numerical experimentation, as exemplified in the work of Shaw and others, is akin both to the testing of a theory and to its theoretical investigation. But in testing, the point of interest is usually how close a particular calculated value is to the observed value, whereas in numerical experimentation it is usually the overall quantitative behavior of the model that is being assessed. And numerical experimentation differs from the customary theoretical investigation in educing the quantitative rather than the verbal or symbolic implications of a set of assumptions. It is shown in the chapters that follow that as meteorological theories have come to be more often expressed in mathematical terms and as more powerful calculating aids have become available, numerical experimentation has become an increasingly important methodology.

In some cases calculations had the effect neither of supporting nor of weakening prevailing theories, but of showing that something unexpected was going on and that a new theory was called for. In a paper of 1870 William Thomson showed that existing tables for the vapor pressure of water were significantly in error when applied to raindrops. These tables, of vapor pressure as a function of temperature, had been calculated on the assumption of a planar surface of water. At a convex surface, as that of a raindrop, evaporation occurs up to a higher partial pressure of water; the greater the curvature, the greater this elevation of requisite pressure. Thomson's calculations were, however, difficult to reconcile with observations indicating that cloud formation often takes place when the relative humidity is only 80 or 90%; the water droplets of the cloud, having a highly curved surface, should readily evaporate even for a relative humidity of 100%. It was not until 1921 that H. Koehler solved this puzzle by showing that the droplets contained salts that had the effect of lowering the vapor pressure of the water (Neis, 1956, pp. 28–29).

A second example of a calculation leading to the discovery of a phenomenon is provided in a paper published by Helmholtz in 1888. He calculated that in the absence of viscosity a ring of air, whose axis coincides with the earth's, moving north or south from the equator would acquire considerable east–west velocity as a result of coming closer to the earth's axis of rotation: motion through 10° of latitude would result in an east–west speed of 14.18 m/s, through 20° a speed of 57.63 m/s and through 30° a speed of 133.65 m/s. Since these speeds are much beyond usual wind speeds, he carried out calculations to see whether viscosity or thermal conductivity might account for the reduction. He found that when air moves in smooth layers, neither effect can cause a significant slowing. Helmholtz then concluded that "the mixing of differently moving strata of air by means of whirls" must be the cause:[11]

> In the interior of such whirls the strata of air originally separate are wound in continually more numerous, and therefore also thinner layers spirally about each other, and therefore by means of the enormously extended surfaces of contact there thus become possible a more rapid interchange of temperature and equalization of their movement by friction. (p. 93)

Helmholtz here provides a vivid picture of a type of turbulence. We will see in Chapter 6 that Lewis Fry Richardson too had his attention drawn to turbulence as a result of trying to deal quantitatively with atmospheric motion.

Calculation sometimes functioned in yet another way to connect theory to measurable phenomena: calculated data sometimes substituted for observational data in testing a theory. In 1903 Frank H. Bigelow tested his model of cyclones and anticyclones against the pattern of pressures and temperatures at three levels: the sea-level plane, the 3500-ft plane, and the 10,000-ft plane. Relatively few measurements of pressures, temperatures, and winds at the higher levels were available, so Bigelow relied on calculated values for all of these. There was, however, no standard way to calculate higher-level winds on the basis of surface data, so Bigelow devised such a procedure. Bigelow claimed that observational data corroborated his procedure, which he regarded a significant contribution to the science:

> That is to say, we may have daily stream lines on the upper planes by computation from surface data, which are as reliable as those which would be obtained from a long series of cloud observations reduced to annual or monthly means. This is a practical conclusion of much value in meteorology. (p. 78)

Another type of calculated data is smoothed or averaged data. Because so many different atmospheric processes affect the measurements, a particular process may not be discernible in the data. The relationship of variables *a, b,* and *c* may be obscured in individual cases by the effects of unknown variables *d, e, f,* . . . . But the average values of *a, b,* and *c* for a great many cases may reveal the relationship because the effects of *d, e, f,* . . . are averaged out. This is the strategy Elias Loomis adopted to test a formula derived by Ferrel that relates the spatial rate-of-change of barometric pressure to wind speed, wind direction, temperature, and other factors.[12] Loomis wrote, "We cannot expect that the formula will be exactly verified in the case of any storm. If however we determine the average values of the elements of a large number of violent storms, the formula ought to represent these average values." Loomis used the data from 81 storms to calculate average values. He concluded that the results "appear to demonstrate that the principles of Ferrel's formula are correct, except that the effect of friction is considerably greater than he has supposed."

Still another way calculation served to test theories was in providing a measure of the fit between theory and data. A striking example of this is Laplace's attempt in 1823 to detect the atmospheric tide caused by the moon, an attempt that also made use of data averaging. Laplace wrote

> It is especially here [with atmospheric tides] that the necessity is felt of employing a very large number of observations, of combining them in the most advantageous manner, and of having a method to determine the probability that the error of the results obtained is confined within narrow limits, a method without which one is liable to present as laws of nature the effects of irregular causes, which has often happened in meteorology (in Sheynin, 1984, p. 58).[13]

Using a record of barometric measurements taken three times daily over an 8-year period, Laplace calculated the probability that the observations indicate a lunar tide. He found this probability not high enough to justify the conclusion that there is a lunar effect. He further calculated that if the actual effect were of the size indicated by the data he used, then nine times as many measurements would be needed to confirm its existence (Stigler, 1986, p. 151).

These, then, are ways calculations have served to connect theory and observations. Since the physics-based theories of the atmosphere have been mathematically complex, the calculations have been complex. Indeed, in many cases the relevant calculation was possible only by employing approximations or some calculating device. Thus in calculating the kinetic energy produced by the vertical motion of a parcel of air, H. Peslin employed a graphical method of calculation and Theodor Reye used an approximation (the first two terms of a binomial expansion) (Kutzbach, 1979, pp. 57, 94). Julius Hann (1883, p. 66) recommended the use of a planimeter for obtaining the total yearly insolation. Bigelow, in the study cited above as in most of his other quantitative studies of the atmosphere, of which there were many, made much use of tables to carry out the calculations. Also in the study by Loomis described above, tables were used in several steps of the calculation.

Since a calculation is an expression of the underlying theory, the number produced by a calculation serves as a test of the theory, when the actual value is known, or as a prediction, when it is not. When the devising of an algorithm is separate from the carrying out of the algorithm, then the latter task can be turned over to people ignorant of the theory or to machines. And when the carrying out of algorithms is thereby expedited, both a theory's vulnerability to disconfirmation and its power to predict are increased.

Chapter 4 A Practical Tradition

Weather Forecasting

Weather signs, such as a small halo surrounding the sun as a portent of rain, have been a part of most cultures from the beginning of history,[1] but astrologers may have been the first professional forecasters. Astro-meteorology in the West goes back at least to Ptolemy's *Tetrabiblos* (*ca.* 160 A.D.), which was the main authority for this practice through the Middle Ages. The invention of printing in the 15th century gave a great boost to astrological weather prediction, principally because of the inclusion of such predictions in almanacs (Shaw, 1926, p. 106). Books on astro-meteorology were published in large numbers until the end of the 18th century; one of the most popular was Giuseppe Toaldo's *Della vera influenza degli astri sulle stagioni e mutazioni di tempo* [On the true influence of the stars on the seasons and changes of weather], published in 1770 (Middleton, 1965, p. 17). In the 19th century that practice almost disappeared, yet by the end of the century weather forecasting was more popular than ever. Willis Moore, Chief of the U.S. Weather Bureau, remarked in 1898, "There is hardly a daily paper that does not publish weather forecasts in a prominent place, and there is scarcely a reader who fails to note the predictions" (p. 12). It was a new technique of forecasting that generated this popularity.

The Weather Map

The new technique was called the synoptic method. Its premise was that knowledge of the present weather over a broad area can yield foreknowledge of the weather at points within that area. Its practice required the construction of same-day weather maps. The synoptic method displaced earlier local methods almost as soon as a means of rapid communication made it technically feasible, and until recently it provided the foundation for virtually all forecasting.

It was the telegraph that made synoptic meteorology feasible. Indeed, in the late 19th century, synoptic meteorology was sometimes known as meteorological telegraphy (Khrgian, 1959, p. 139). The practicality of the telegraph for long-distance communication was first demonstrated by Samuel Morse's Baltimore-to-Washington line in 1844, and in the 1850s the telegraph was used, by James Glaisher in England and Joseph Henry in the United States, to construct same-day weather maps (Fleming, 1990, pp. 141–142). In 1863 the French national weather service began issuing daily weather maps. Next to do so was the U.S. weather service in 1871, followed within 6 years by the British, the Russian, the Danish, the Swedish, the Belgian, the German, the Algerian, the Australian, and the Austrian (Shaw, 1926, p. 287). Although climatic maps—such as maps of prevailing winds or average temperatures—were, as we saw in Chapter 2, fairly common by mid century, it was not until these national weather services began issuing daily maps that maps of weather conditions at a particular time became common.[2]

Weather maps revealed regularities that were not apparent in climatic maps. Buys-Ballot's law was thus discovered. It was found that there is generally a counterclockwise flow of air about the center of a region of low pressure, that wind speeds tend to be greater in the southern part of a low-pressure area, and that temperatures tend to be lower in the western and northern parts. Especially cold weather, it was found, usually occurs in regions of high pressure and most rain, in regions of low pressure. The usual courses of storms were charted. And weather maps made clear the general tendency, in the north temperate zone, for atmospheric conditions to move from west to east.

These, and many other, regularities contributed to meteorologists' ability to forecast the weather. The last-mentioned regularity, however, was of paramount importance. David Brunt has written

> The forecasts of the first 50 years of the [British] Meteorological Office were largely based on the assumption that weather travels in a general west–east direction, and that any depression will continue to move along the path it has followed during the past 6 or 12 hours. (1951, p. 120)

And according to a 1918 report of the National Research Council prepared by the U.S. Weather Bureau, "The art of weather forecasting rests almost entirely upon the fundamental proposition that weather travels" (p. 122).

At the turn of the century, forecasts were generated in the following manner. Each day forecasters constructed several types of synoptic chart using telegraphic reports from a hundred or more locations.[3] The most important was a map showing the reported barometric pressures, on which isobars were drawn. In addition, barometric-change maps were usually drawn that showed 12- or 24-h changes. Other maps, showing temperature, temperature change, precipitation, humidity, wind, or cloud cover, were also drawn. Figure 1 gives an example of a weather map used in forecasting.

It was above all the pattern of isobars to which forecasters paid attention, since certain arrangements of isobars were thought to be associated with particular types of weather. The main task was to form a mental or physical picture of the pressure distribution for the coming day. The forecasters often simply assumed that a low-pressure area would continue its observed motion, in both direction and speed; they sometimes made use of maps showing the usual tracks of lows and the average speed of a low along each track.[4] The speed was used to estimate the amount of precipitation, on the assumption that the slower the motion the more precipitation. In addition, Buys-Ballot's law, which yielded wind direction and speed from knowledge of the pressure distribution, was regularly used. Such information and

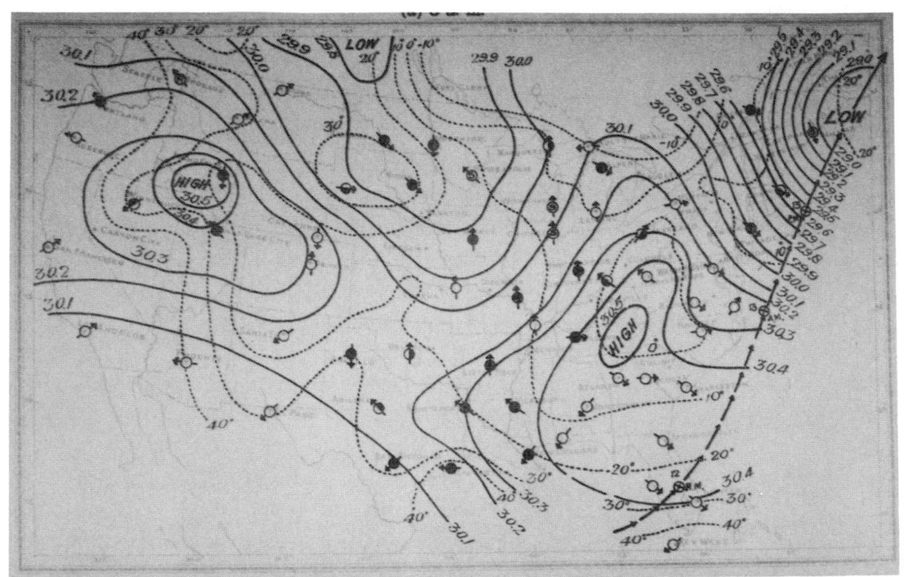

Figure 1 A map of meteorological conditions in the United States at 8 a.m. on 14 February 1899 (*Monthly Weather Review*, Vol. 27, plates following p. 88).

considerations led to a forecast, which was a verbal description of the coming weather, usually no more precise than "rainy and windy" or "clear and cold."

The synoptic method, with its exclusive reliance on mapped information, was by the end of the 19th century in use throughout Europe and in the United States. J. M. Pernter of the Austrian weather service wrote in 1903, "The knowledge of the weather conditions for every place and for every type of pressure distribution offers the only entirely satisfactory empirical basis for weather predictions . . ." (p. 160), and a 1916 publication of the U.S. Weather Bureau reported, "After an experience of many years, the forecasters of the Weather Bureau continue to make all forecasts of every character as to future weather conditions solely on the basis of synoptic weather maps" (p. 69).

Thus the weather map gave forecasters a method for making predictions that did not depend on a theoretical understanding of the atmosphere. Pernter wrote,

> Since we have to do only with theorems founded entirely upon experience, the persons best qualified to make the predictions are those who through long years of practice have collected the most theorems as to the variations in the forms of pressure distribution, and have also learned by practice the many modifications to which these theorems are subject. (1903, p. 161)

The considerable achievements of theoretical meteorologists, described in Chapter 3, played almost no part in forecasting. Indeed, Napier Shaw claimed that "the introduction of the weather-map led to a curious alienation of the experimental and theoretical physicists from the study of weather" (1926, p. 154). He explained that

. . . the compilers of observations and maps were profoundly conscious that the experiments and theories of the physical laboratories offered no real explanation even of the broadest features of the distribution of pressure and temperature, and that to plunge into the study of minute details, mathematical and physical, when the outlines were an unsolved riddle was equally a waste of time and energy. So there came about a sharp division, physicists on the one side, regarding the efforts of the observers and map-makers as quite unscientific and sometimes suggesting that competent mathematicians should be invited to take the matter up; and meteorologists on the other side, equally firmly convinced that to invite the mathematicians to solve a problem which they could not specify was the same sort of mistake as inviting Newton to solve the problem of the solar system without the previous assistance of Kepler's laws.

The method did, of course, depend on data gathering, but for the most part only on the data used in drawing the latest maps; the vast accumulation of earlier observations, which was the basis of climatology, was seldom used in any explicit way. Thus the general picture is of climatologist, theorist, and forecaster engaged in quite separate activities.

Skepticism about Weather Forecasting

Side by side with the tradition of weather prognostication has always moved a restraining, skeptical tradition. Early in the 19th century the astronomer François Arago declared that no one who had scientific character to lose would prophesy weather (Scott, 1873, p. 378), and when state-sponsored forecasting became common later in the century, the skeptics became more vocal. Robert Fitzroy, remembered today as commander of the *Beagle* during the famous expedition for which Charles Darwin was naturalist, became the first director of the British Meteorological Office in 1855 and began issuing weather forecasts in 1861.[5] The harsh criticism his forecasting excited, especially from scientists, was very likely a contributing cause of his suicide in 1865 (Hughes, 1988, p. 201). In the following year a committee of the Royal Society recommended that daily forecasting be stopped for the reason that it was not based on scientific knowledge, and not until 1879 did the Meteorological Office resume the practice.[6] The Danish weather service, established in 1872, was for many years extremely cautious in this respect, restricting itself to three forecasts, "fine weather," "unstable weather," and "bad weather," and the Swedish weather service, established just a year later, refrained altogether from making forecasts until 1905 (Khrgian, 1959, p. 156). According to David Brunt, "The use of weather maps for such purposes [that is, forecasting] was regarded in scientific circles with suspicion, and was described as 'empirical,' a word which, in the mouths of scientific men, is a heavy missile" (1951, p. 117). In the second half of the 19th century the scientific critics were perhaps most numerous in England but included Julius Hann in Austria, Gustav Hellman in Germany, and J. G. Galle and Georges Rayet in France.

Both proponents and opponents of forecasting tried to prove their case by measuring the accuracy of forecasts. The vagueness of the official forecasts, which with a few exceptions were entirely nonquantitative in the period up to World War I, aided the proponents. Typical is an evaluation, made in 1883 by the director R. H. Scott, of the forecasts of the British Meteorological Office. He found that 35% were "quite correct" and an additional 41% were "more than half correct," which for him bespoke a success rate of 76% (Scott, 1883, p. 72). In 1903 the Austrian weather service claimed a success rate of just above 80% (Pernter, 1903, p. 162). As representative of the opponents we may take A. Mallock, who, writing in *Nature* in 1914, compared the daily weather forecasts for London for all of 1913 with the actual weather. He concluded that someone following the rule "To-morrow will be like to-day" would have been right almost as often as the Meteorological Office was.[7]

On the whole, however, the proponents of forecasting had the better of it even in scientific circles, and they certainly had the overwhelming support of the general public. In 1883 R. H. Scott wrote that daily forecasting "has really been forced upon meteorologists by the demand of the public to see in the newspapers some statement as to probable weather" (p. 69). In the 20th century the skeptics have become fewer and quieter, yet as late as 1959 a leading meteorologist, Tor Bergeron, wrote that public weather forecasting had started a hundred years too soon, in 1860 instead of 1960 (p. 442).

"Science, Not Art"

The modest success of the forecasts partly explains the skepticism many scientists felt toward them. More important though was the common perception of forecasting as an unsystematic, judgmental process that was not based on scientific knowledge. This perception was fairly accurate. The meteorologist Richard Reed wrote that "physical principles and theoretical concepts played little, if any, role in practical weather prediction up to the time of the First World War" (1977, p. 391). Willis Moore wrote in 1898

> No exact rule in regard to them [low-pressure areas] can be laid down. Empirical reasoning, and intimate association with the charts, day after day and year after year, in the main equip the successful forecaster for his important functions. (p. 13)

In the 1918 report of the National Research Council cited above, we read, "Practically all of the rules known and used in the art [of forecasting] have been established empirically; some of them have been formulated, but in a considerable proportion of cases the rules which govern the forecaster are exercised subconsciously" (p. 122). Here, and elsewhere, the necessity of long experience in forecasting and the impossibility of reducing the practice to rules were asserted. Here, and elsewhere, forecasting was classified as an art rather than a science.

Not many people, however, were comfortable with this state of affairs. The unease might have been ameliorated by clear improvement in forecasting technique or in accuracy, but this was not forthcoming. In fact from 1870 to 1920 there was neither much change in the way forecasting was done (Bergeron, 1959, p. 449;

Table I
Weather Indications along the Scale of an English Barometer Made in about 1700[a]

Summer	Winter
Very Dry	Hard Frost
Settld Fair	Settld Frost
Fair	Frost
Changeable	Uncertain
Rain	Snow
Much Rain	Much Snow
Stormy	Tempest

[a]Goodson (1977, p. 51).

Brunt, 1951, p. 120; U.S. Weather Bureau, 1916, p. 69; Whitnah, 1961, p. 63) nor any discernible increase in accuracy (Douglas, 1952, p. 16; Reed, 1977, p. 395). This stasis in forecasting contrasted with the striking progress in the late 19th and the early 20th century in climatology and in dynamical meteorology.

The unease contributed to the attraction of "science, not art," that is, the attraction of fully specified methods in forecasting. This attraction manifested itself in different ways. Some meteorologists wanted to use theories of physics in making predictions. Some sought an independent theory. Some, especially the statistically inclined, thought that a careful study of the observational data could yield forecasting rules.

A few of the earlier forecasting techniques had in fact been fully algorithmic, such as the use of the barometer as a "weather glass" (as illustrated in Table I) and—a much more complex algorithm—the astro-meteorological method presented in Leonard and Thomas Digges's *Prognostication everlastinge* (1576).

But these techniques had long since been discredited, and from the beginning of state weather-forecasting until the 1950s almost all forecasts were arrived at by subjective processes that involved little calculation. In the late 19th and early 20th century there were numerous attempts to systematize weather forecasting.

One of the most successful, at least in point of winning adherents, was that of the Scottish meteorologist Ralph Abercromby. Abercromby's method, explained to a lay audience in *Weather* (1887, pp. 25, 54), was based on a classification of the forms that isobars assume on a weather map. He held that there were exactly seven well-defined configurations, shown in Figure 2 (top diagram): cyclone, secondary cyclone, V-shaped depression, anticyclone, wedge, col, and straight isobars. The weather associated with each of these configurations was described verbally and in separate diagrams, one of which is shown in Figure 2 (bottom diagram).

It was common in the late 19th and early 20th centuries for meteorologists to provide lists of rules for arriving at a forecast. Examples are the formal listing of rules for tornado prediction given in 1884 by the U.S. meteorologist John Finley and the rules contained in the 1904 article "Attempts at methodical forecasting of the weather" by the French meteorologist Louis Besson (Galway, 1985, p. 1506).

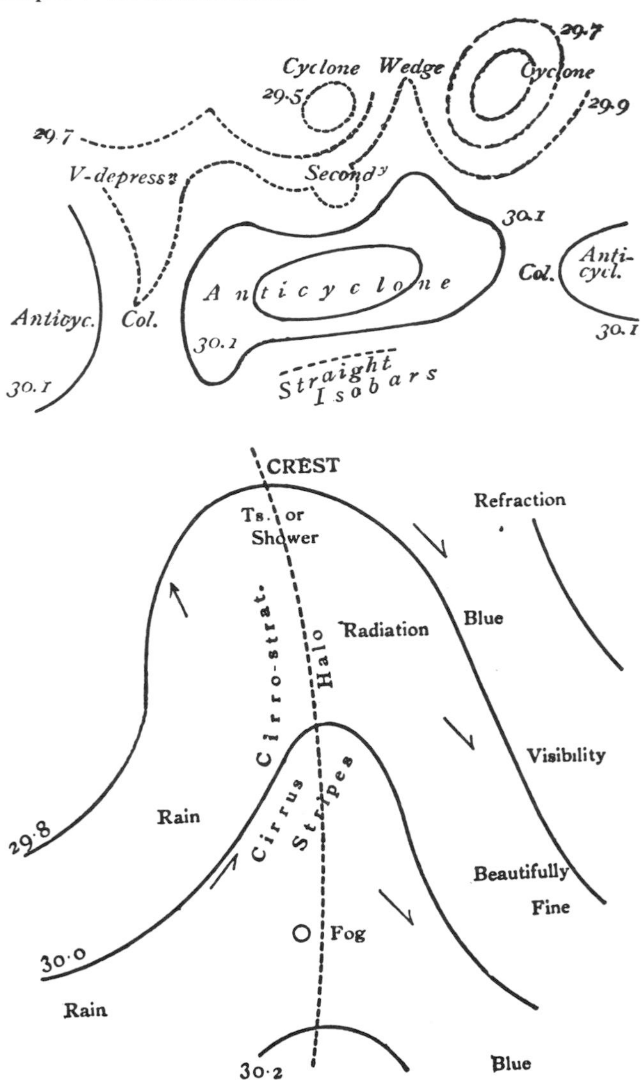

Figure 2 The top diagram shows the seven configurations of isobars according to Abercromby's *Weather* (1887, pp. 25, 54). The bottom diagram shows the weather associated with a wedge.

In 1905 another French meteorologist, Gabriel Guilbert, won first prize in an international competition of forecasting methods (Fassig, 1907, p. 210). Guilbert, who presented his method as a series of rules, wrote that weather forecasting was "empirical up to the present time, without strict rules, and based upon an incommunicable personal experience. . . ." (1907, p. 212). These methods—Abercromby's, Finley's, Besson's, and Guilbert's—like almost all others, were not fully algorithmic even though they could be expressed as a set of rules, since there was much judgment involved in the application of individual rules and in the choice of which rules to apply.

Calculation in Weather Forecasting

It must be remembered that all of these systems of forecasting began with the drawing of the current weather map, and although subsequent steps seldom involved calculation the first step required a great deal of calculation. The tasks mentioned earlier—converting units, applying corrections, calculating the values measured indirectly, and reducing to standard conditions—had to be done by the forecasters as well as by the climatologists.[8] There was, however, this great difference: the forecasters needed the processed data within a few hours of the observations.

The use of numerical tables indeed speeded computation immensely, and the climatologists and the theoreticians seem to have been satisfied with the method, at least in those situations where the relevant tables had already been computed. But for the forecasters the turning of pages and the looking down columns and across rows could take too long. Thus it was especially the forecasters who favored a single international system of units to eliminate all the conversion of units (Gold, 1934, p. 124). And it was especially the forecasters who were interested in graphical procedures of calculation and the use of slide rules, which were somewhat faster than the use of tables.

Another graphical technique, nomography, began to be used. A nomogram is a figure presenting a quantitative law in such a way that the implication of the law, in any particular case, is readily determinable, usually by seeing where a straightedge, placed so that it connects points on two scales, cuts a third scale (Figure 1 of Chapter 8 is an example). For example, in 1906 John Ball published, in the *Quarterly Journal of the Royal Meteorological Society,* a nomogram for calculating dew point and vapor pressure from wet- and dry-bulb thermometer readings; thus a single diagram might serve in place of the 33-page table, part of one page of which was shown in Figure 1 of Chapter 2. Most of the development of special-purpose slide rules and of graphical techniques took place after World War I and is therefore treated more fully in Chapter 8.

The fact that weather conditions at any particular time could be satisfactorily described by a weather map suggested a procedure of forecasting that seemed to bypass theory altogether: find a past weather map that closely resembles the present map and assume that the weather will now change as it did on that earlier occasion.[9] This is, in the words of the meteorologist Lewis Fry Richardson, to use the history of the atmosphere "as a full-scale working model of its present self" (1922, p. xi). This method has been tried repeatedly in the last hundred years, the increasing archive of past weather maps and new schemes for identifying similar maps always serving to renew optimism. In the years around 1900 the U.S. Weather Bureau did a good deal of work on such methods,[10] and in World War I Ernest Gold compiled an *Index of Weather Maps.* Chapter 9 describes a large-scale effort to forecast in this way that was made during World War II. Despite its perennial appeal and the many attempts, this method has never achieved clear success or become widely used.

Frequently meteorologists sought to find a statistical regularity that would be of use in forecasting. Thus in 1876 Clement Leh found that on average the path of a low-pressure area is at right angles to the direction of greatest pressure change, and in 1879 P. I. Brounov found that the path makes an angle of 28° with the isotherms (Khrgian, 1959, p. 177). Many statistical studies of the angle between wind direction and barometric gradient were carried out. The studies of storm tracks yielded maps showing the most common tracks, and such maps were widely used in forecasting (Khrgian, 1959, p. 148; National Research Council, 1918, p. 124; Whitnah, 1961, p. 227). There were many attempts to identify, by statistical analysis, weather types associated with the different tracks and with different times of the year (Khrgian, 1959, pp. 175–180; National Research Council, 1918, p. 123; Scott, 1873, p. 384).

Some meteorologists formulated probabilistic rules, such as

> If we take the area from Valencia to Helder, and from Nairn to Rochefort, we find that whenever the difference of barometrical readings between any two stations is 0.6 in. on any morning, the chance is 7 : 3 that there will be a storm before next morning . . . somewhere within the area covered by our network of stations. (Scott, 1869, p. 340)

Probabilities were computed for wind speeds, temperatures, precipitation, fog, and cloudiness in different parts of a cyclone.

In 1870 Wladimir Köppen published "On the sequence of the non-periodic variations of weather, investigated according to the laws of probability." His main conclusion was that each weather type tends to persist. He found, for example, that if it has rained in Brussels for 10 days, then on the 11th day there will be rain in four cases out of five (in Scott, 1873, p. 379). Köppen found also that the longer a weather type has lasted, the greater the probability it will persist another day. So the probability of cold weather on the day following 2 months of cold weather is greater than on the day following just 5 days of cold weather. Köppen's result—that one can expect the present weather to continue, the more surely the longer it has lasted—hardly made weather records any more valuable to forecasters. This was, indeed, the result in almost all the statistical studies: they provided general guidance to forecasters but no specific predictions that could be relied upon. Yet studies of this sort were an important and continuing link between the empirical and the practical traditions.

By the turn of the century meteorology was established as a discipline. Its status as an empirical science resulted mainly from the work of climatologists. Its status as a theoretical science was achieved almost entirely by applying physics to atmospheric phenomena. An institutional base was furnished by governments largely because of meteorologists' ability to forecast the weather. So all three traditions provided vital elements to the new discipline, yet the traditions remained fairly independent of one another. Within each of them calculation had important but not leading roles. In the first decade of the 20th century a program was enunciated to unite them through intense calculation.

Part II | Meteorology in the First Half of the 20th Century

. . . a mighty problem looms before us and we can no longer disregard it. We must apply the equations of theoretical physics not to ideal cases only, but to the actual existing atmospheric conditions as they are revealed by modern observation. These equations contain the laws according to which subsequent atmospheric conditions develop from those that precede them. It is for us to discover a method of practically utilizing the knowledge contained in the equations.

VILHELM BJERKNES, 1914

The world's population increased from $1\frac{1}{2}$ billion in 1900 to $2\frac{1}{2}$ billion in 1950. In the same period the number of scientists grew at a faster rate, and the number of meteorologists at a faster rate still. The new positions resulted from the establishment of meteorology as an academic discipline, together with a vast increase in the size and number of universities; the expansion of state weather services, both civilian and military; and a new demand for meteorological expertise by airlines, shippers, and other businesses. The great increase in the number of meteorologists was accompanied by great changes in meteorological observation, theory, and practice, many of them resulting from a determination to use the laws of physics to unify the meteorological traditions.

Chapter 5 | Vilhelm Bjerknes's Program to Unify Meteorology

In the years around 1900, Otto Lilienthal, Ferdinand von Zeppelin, Samuel Langley, Wilbur and Orville Wright, Louis Blériot, and others were working to make air transportation practical. At the same time another scattered group of enthusiasts—A. L. Rotch in the United States, H. H. Hildebrandsson in Sweden, Léon Teisserenc de Bort in France, Richard Assmann in Germany, W. H. Dines in England, and others—were also intent on reaching up into the atmosphere. These men were meteorologists whose successful efforts led to the recognition of a new branch of the science.

Aerology

For many purposes, it is not unreasonable to treat the atmosphere as two-dimensional. Meteorologists do so in using most weather maps. The principal justification is that the volume of air meteorologists typically analyze is as thin, relatively, as a playing card: most clouds and storms occur entirely in the lowest 10 km of the atmosphere, while they may move halfway across a continent in a day. Until the last decade of the 19th century, there was the additional justification that almost all meteorological measurements were of atmospheric conditions at the earth's surface.

There had long been attempts to gather upper-air data. Some information, of course, could be obtained by placing instruments high up on mountains, and beginning in the 1870s meteorologists maintained a number of mountain observatories. For carrying instruments above the earth's surface there were balloons and kites. Indeed, some of the earliest balloon ascents were made with meteorological instruments, and in the 1860s James Glaisher in England made 28 ascents to measure the dependence on height of temperature, humidity, and electrical potential (McAdie, 1917, pp. 7–19). But until the 1890s there were only isolated efforts of this sort. Moreover, most of the measurements attained with balloons and kites were unreliable because of the "radiation error" (the heating of the thermometer by direct or scattered sunlight), inaccurate height determinations, or other problems (Khrgian, 1959, pp. 269–275).

In the last decade of the century, systematic measurement of the free atmosphere began with kites, captive balloons, sounding balloons, cloud observations, and pilot balloons. In 1894 A. L. Rotch initiated regular kite observations at the Blue Hill Observatory near Boston, and in the years that followed many European meteorologists—notably Léon Teisserenc de Bort at his own observatory near Paris—made kite observations. The meteorological use of unmanned balloons, which, like the kites, carried light-weight recording instruments aloft, began about the same time. Captive balloons were often used as substitutes for kites at times of low winds. Sounding balloons, which reached much greater heights, were more important to meteorologists. International agreement to launch sounding balloons on specified days ("international balloon days") began in 1896 and lasted until the first world war. Also beginning in 1896 was an international program, led by H. H. Hildebrandsson, to measure the heights and motions of clouds. And in the years around 1900 several meteorologists used theodolites to track small balloons—called "pilot balloons" from the aeronaut's practice of releasing such a balloon, prior to his own ascent, to see how the winds would carry him—in order to measure air movements.

The new data proved interesting and useful. Most surprising and most significant was the discovery of the stratosphere. In 1902 Teisserenc de Bort reported, on the basis of 236 ascents of sounding balloons, that there was an atmospheric zone, whose variable lower boundary was at a height of 8 to 12 km, where temperature was fairly constant as a function of height. In the same year Richard Assmann in Berlin published results that corroborated this finding, and in 1908 Teisserenc de Bort suggested the names "troposphere" and "stratosphere" for the zone adjacent to the earth's surface and the isothermal zone, respectively (Schneider-Carius, 1955, pp. 308–310).

At a meeting of the International Commission for Scientific Aeronautics in 1906 Wladimir Köppen proposed the name "aerology" for the study of the upper atmosphere. Köppen, who regarded aerology, like climatology, as a branch of meteorology, described it as a threefold science: the science of measurement in the free atmosphere, the collection of data, and the relating of data to physical law (Neis, 1956, p. 19). This extension of meteorology, besides providing new data and raising new questions, was a stimulus to meteorological research generally (Kutzbach, 1979, p. 186).

In the opening address to the International Meteorological Congress in 1900, E. Mascart said,

> . . . we see how important must be the study of the upper regions of the atmosphere, of its temperatures and its winds, and of the distribution and transformation of the clouds. . . . We are on the way to realize the complete scientific conquest of the atmosphere. We shall perhaps soon see aerial stations furnishing daily reports adapted to improve the present imperfect service of weather predictions. (1901, p. 265)

As this quotation suggests, the new success in studying the free atmosphere, which added the vertical dimension to meteorological measurement and analysis, made meteorologists optimistic about their science. In 1901 Frank H. Bigelow said

> It is not too much to say that meteorology is turning over a new leaf in these days. Speculative theories are being discarded, and we must build anew upon the motions

> of the air as disclosed by cloud observations, and upon the gradients of pressure, temperature, and vapor tension as measured in kite and balloon ascensions or by cloud computations. (1902, p. 22)

In 1905 Julius Hann characterized the years around 1900 as a transition period for meteorology, citing as evidence new programs to obtain upper-air measurements (1906, pp. v–vi). Perhaps the most optimistic of all was the Norwegian physicist-turned-meteorologist, Vilhelm Bjerknes.

From Physics to Meteorology

At the turn of the century there were a number of people who thought that the time had come to use the laws of physics to forecast the weather. This was the view of Cleveland Abbe in the United States; in 1890 his *Preparatory Studies for Deductive Methods in Storm and Weather Prediction* had been published by the Signal Office. This was the view of Napier Shaw; it was his aim, according to Ernest Gold, to replace empirical forecasting with "methods depending on the direct application of physical principles" (1945, p. 216). For a time this was the view of the Austrian Felix Exner, who in the years 1906 to 1910 tried to develop predictive methods based on physics (Platzman, 1967, p. 539).[1] But by far the most influential in this respect was the Norwegian scientist Vilhelm Bjerknes.

In 1903 Bjerknes began proselytizing on behalf of the new creed: that weather forecasting should become an exact science and that the only way of achieving this was by building up from the laws of physics. Bjerknes, unlike Abbe, Shaw, and Exner, believed so strongly in the realizability of physics-based forecasting that he directed all his efforts toward it. Moreover, Bjerknes convinced others to do the same, and by 1920 Bjerknes's creed was shared by many.

Bjerknes had been trained as a physicist, mainly at the University of Kristiania (now the University of Oslo), receiving an M.S. in 1888 and a Ph.D. in 1892. He had studied also in Paris, where he attended Henri Poincaré's lectures, and in Bonn, where he worked for 2 years with Heinrich Hertz. Bjerknes held various university positions in mathematical physics in Sweden and Norway until 1913, when he became director of a newly established institute of geophysics in Leipzig. In 1917 he returned to Norway to start a geophysical institute in Bergen. There he attracted talented collaborators and established what is known as the Bergen School of Meteorology.

Bjerknes's early work was in electrodynamics and hydrodynamics. In the 1890s he succeeded in proving two important theorems, so-called "circulation theorems," that were generalizations of theorems on fluid motion proved by William Thomson and Hermann von Helmholtz. This was a time when many physicists, including Carl Anton Bjerknes, Vilhelm's father, were studying vortex motion mathematically. Bjerknes greatly extended the range of applicability of such theorems by dispensing with the assumption that the density of the fluid is a function of pressure only. Soon thereafter he became interested in meteorology and hydrography, apparently because he saw them as areas where his theorems might

Table I

The Seven Variables That for Bjerknes Characterized the Atmosphere[a]

v_E = eastward component of the velocity of air	p = pressure
v_N = northward component of the velocity of air	θ = temperature
v_H = upward component of the velocity of air	ρ = mass density
	μ = water content per unit volume

[a] Each of these may be regarded as a function of four independent variables—three variables specifying position and one variable specifying time.

be fruitfully applied (Friedman, 1982, p. 345; Bergeron, 1959, p. 452; Pihl, 1970, p. 168).

What was most striking about Bjerknes was his optimistic view of what meteorology had achieved and would achieve. Bjerknes thought that observational meteorology had already reached its main goal: the ability to give a complete characterization of the atmosphere at one place and at one time. Meteorologists could do this because they could measure, even high in the atmosphere, the seven quantities that are listed in Table I. Thus in 1913 he asserted that "observational meteorology has practically completed and perfected her task, no matter what future advances may yet be made in instruments and in organization" (Bjerknes, 1914, p. 13).

Bjerknes seems to have thought that any property of the atmosphere either could be defined in terms of these seven variables, as the force exerted by the wind or the vapor pressure of water could be, or was not significant to meteorology, as the smell of the air and the aurora borealis were not. Thus Bjerknes was, in a sense, providing a definition of the science. For the meteorologists who accepted his view, as most seem to have done, atmospheric phenomena not describable in terms of these seven variables were not part of the science. Indeed, optical effects in the atmosphere, photochemical reactions in the atmosphere, and the electromagnetic phenomena of the ionosphere are matters that have been investigated more by physicists and chemists than by meteorologists.

The Program to Calculate the Weather

Bjerknes claimed much more. Not only had observational meteorology reached its main goal, theoretical meteorology had too. The important laws governing the atmosphere were already known: the principles of hydrodynamics and thermodynamics sufficed to determine completely the values of the seven variables. As Bjerknes *et al.* put it

> Inasmuch as we know the laws of hydrodynamics and thermodynamics, we know the intrinsic laws according to which the subsequent states develop out of the pre-

ceding ones. We are therefore entitled to consider the ultimate problem of meteorological and hydrographic science, that of the precalculation of future states, as one of which we already possess the *implicit* solution, and we have full reason to believe that we shall succeed in making this solution an *explicit* one according as we succeed in finding the methods of making full practical use of the laws of hydrodynamics and thermodynamics. (1911, p. 4, emphases in the original)

The relevant principles could be expressed as a set of seven equations, equations that had been discovered in the preceding two centuries. Newtonian mechanics, as presented in the *Principia* (1687), applied only to bodies, such as homogeneous spheres, that could be treated as point masses. A. C. Clairaut made hydrostatics a Newtonian science; Leonhard Euler did the same for hydrodynamics. This gave four equations: force equals mass times acceleration in each of three directions, and the mass-conservation equation. These equations were enriched by L. M. H. Navier in about 1820 by adding terms to account for internal friction, and made to apply to motion in a rotating reference frame by G. G. Coriolis in 1835. The work of Robert Boyle, J. L. Gay-Lussac, and William Thomson led to the gas equation, which states that the product of pressure and volume is proportional to the absolute temperature. The sixth equation, the hydrostatic equation, first appeared in the work of Laplace around 1800. Thermodynamics, developed especially by Hermann Helmholtz, Thomson, and S. S. Clausius, led to the final equation: the energy-conservation equation. The seven resulting equations form a mathematically complete set.[2]

Bjerknes was not original in regarding the problem of the weather as having been solved in principle by physics. In 1830s William Whewell omitted meteorology from his list of inductive sciences because he believed it to be essentially the application of the laws of physics to the atmosphere (Shaw, 1926, p. 117). In 1851 the Russian meteorologist M. F. Spasskii wrote that "the basis of meteorological theory . . . is as definite as the basis of astronomical theory" and that the problem of weather forecasting can be posed as a mathematical problem (a prerequisite to the solution being, however, knowledge of the previous state of the weather throughout the world) (Khrgian, 1959, p. 233). And Cleveland Abbe had in 1901 even set down the same system of equations as Bjerknes did (Saltzman, 1967, p. 590). To Bjerknes, however, belongs credit for undertaking a vigorous program to achieve physics-based forecasting and for inspiring others to work along similar lines.

Bjerknes's creed, though optimistic about what had been done—"we have all the means in our hands to solve these problems" (Bjerknes, 1906, p. 154)—gave no comfort to a meteorologist inclined to relax. Observational meteorology knew how to do its task, but to carry out that task would require a much greater commitment of resources and a higher degree of international cooperation. Theoretical meteorology had a secure foundation, but little had as yet been built upon it, and to build high enough to explain atmospheric phenomena required the utmost exertion. And still there remained the great task of meteorology: to bring together the full range of observation and the full range of theory in order to predict tomorrow's weather. Bjerknes wrote, "What is it that I really seek? Whither am I steering? I could not free myself from the thought that 'There is after all but one problem worth attacking, viz, the precalculation of future conditions.'" (1914, p. 14). But a concentration on calculating future states of the atmosphere would not, Bjerknes thought, narrow the range of meteorological activity since this problem "encompasses all the others" (in Khrgian, 1959, p. 232).

The Graphical Calculus

Bjerknes never believed that the equations describing the atmosphere could be solved analytically: "Obviously the usual mathematical methods will not be adapted to a problem of this sort. There can be no thought of an analytical presentation of the observational results with a subsequent analytical integration of the equations" (Bjerknes, 1914, p. 14). And it may be that until he learned in the early 1920s of Lewis Fry Richardson's work (which is the subject of the next chapter) he did not seriously consider the possibility of doing the calculation numerically.[3] The method Bjerknes proposed was a graphical one: "As the observations are presented by means of charts, therefore all mathematical computations must be recast into graphical operations by means of maps" (1914, p. 14).[4]

This method, which became known as the graphical calculus, was advocated by Bjerknes almost from the beginning of his interest in meteorology. In a course of lectures he gave in 1905 he said

> Suppose that we have two systems of maps, one representing the field of motion at a given time, and the corresponding one representing the field of force upon which the change of motion depends. Evidently, it will then be possible to construct maps representing the field of motion at some later time. . . . (Bjerknes, 1906, p. 153)

Hertz, with whom Bjerknes had studied, had introduced in 1884 a graphical procedure for studying adiabatic processes in the atmosphere. Graphical addition and subtraction were described in Adolf Sprung's widely used *Lehrbuch der Meteorologie* of 1888, and at about the same time Wladimir Köppen and Max Möller used graphical procedures to construct upper-air weather charts (Kutzbach, 1979, pp. 152, 237).

In the first decade of the new century, Bjerknes and his collaborators pursued the development of the graphical calculus, and in 1910 and 1911 they published the first two volumes of *Dynamical Meteorology and Hydrography* (Bjerknes and Sandström, 1910; Bjerknes *et al.,* 1911). These volumes present a great many graphical techniques, such as for multiplying scalar fields, for taking the gradient of a scalar field, and for taking the divergence or the curl of a vector field. An example of the graphical calculus applied to meteorology is shown in Figures 1, 2, and 3, taken from Bjerknes, Hesselberg, and Devik's *Kinematics* (1911).

Bjerknes starts with a map (Figure 1) of India on which wind directions at ground level are indicated by arrows and wind speeds by numbers placed beside the arrows. Using the arrows as guides, he draws a set of lines, the streamlines, to represent the horizontal flow of air (Figure 2). And on the same map, using the numbers indicating wind speed, he draws a set of isolines for wind speed (so that wind speed is constant at all points on a single line).

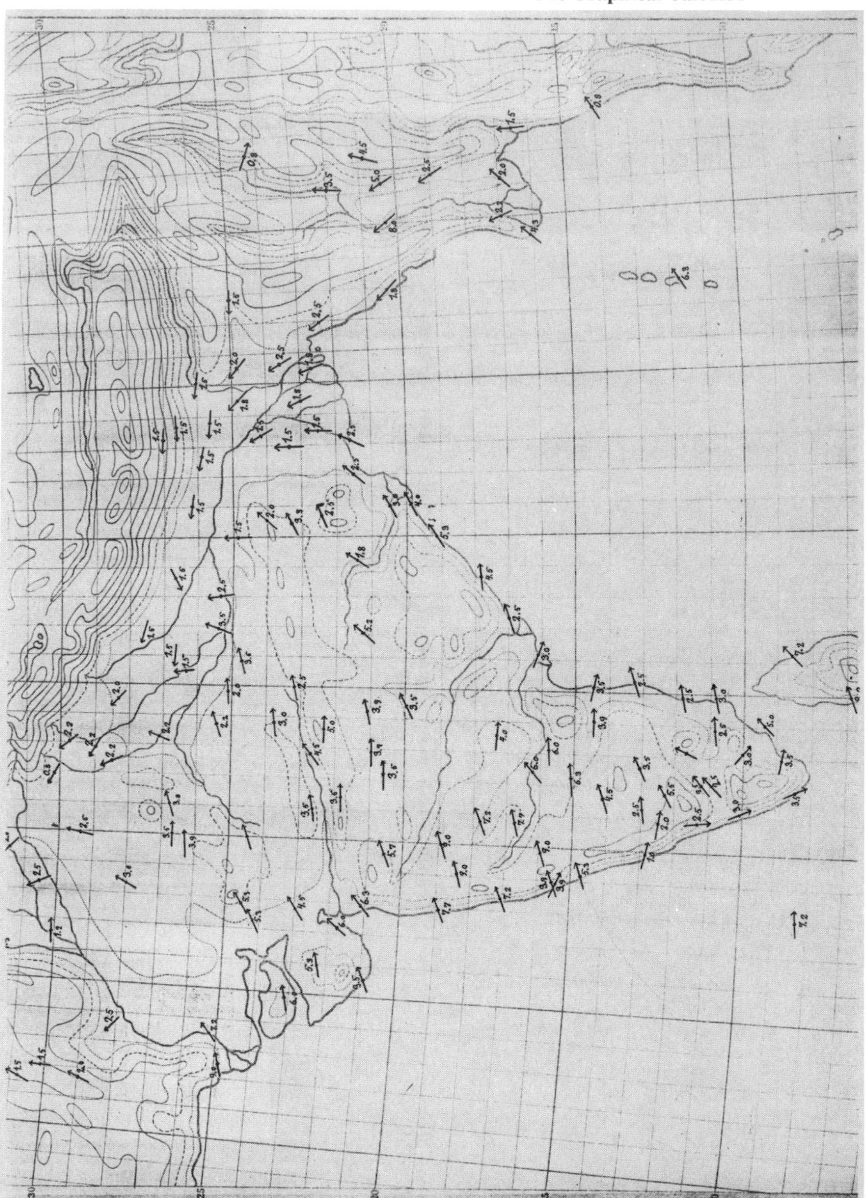

Figure 1 This map of India presents the observational information: the arrows indicate wind direction, and the numbers indicate wind speed in meters per second. (The inclusion of contour lines on the map allows one to see the effect of topography on winds. For example, as Bjerknes points out, the high mountains near the west coast of the southern peninsula deflect the winds southward. This effect is more easily seen in Figure 2.)

Finally he constructs a map (Figure 3) showing the vertical motion of the air at ground level. It is clear that if a thin region just above the ground has a greater horizontal inflow than horizontal outflow, there will be a forced upward motion of

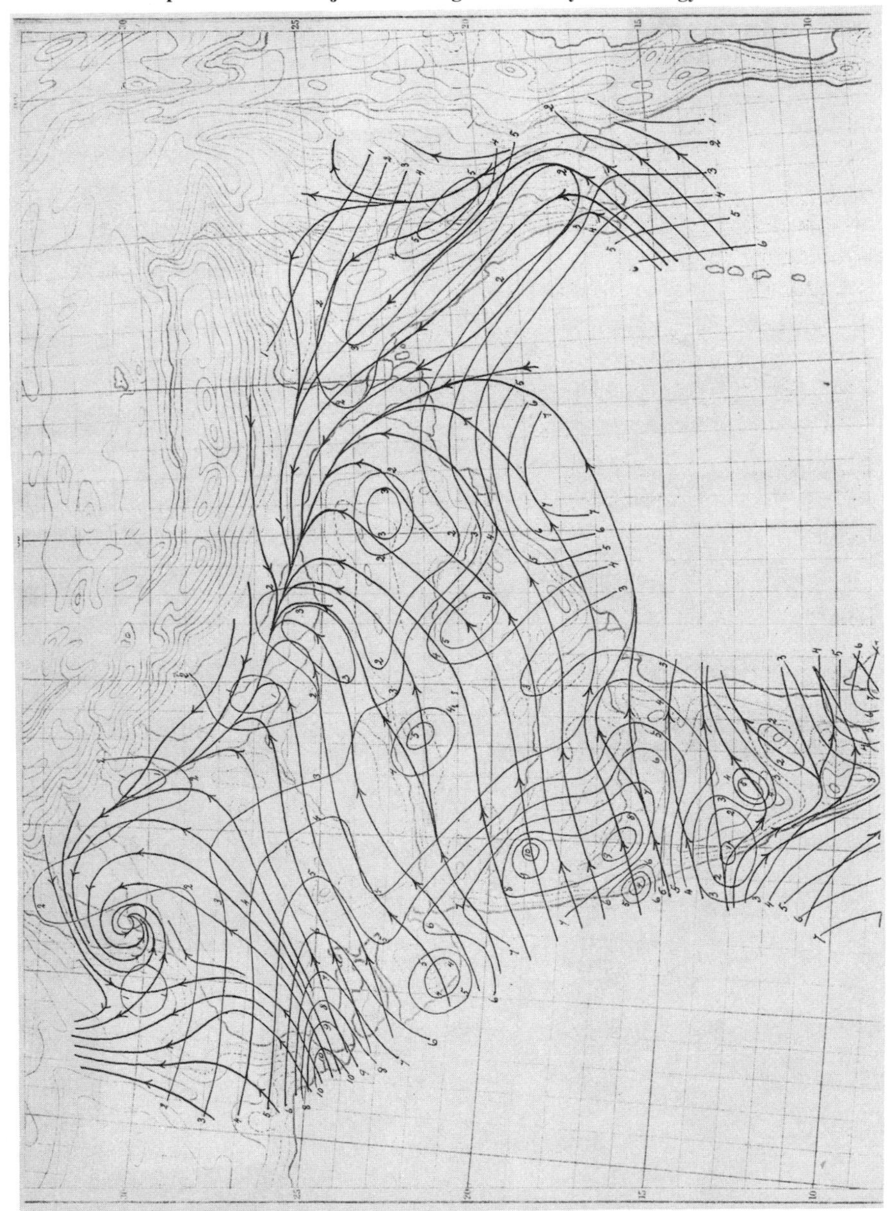

Figure 2 The information on the preceding map was used to construct this map. The thick lines (streamlines) show the direction of the flow of air, the thin lines show the speed of the flow.

air in that region. Bjerknes used graphical differentiation and graphical algebra to calculate the forced vertical motion at each point, and the result is shown in Figure 3, where areas of ascending air are shaded, areas of descending air unshaded.

Figure 3 Graphical differentiation and graphical algebra were used to construct this map from the preceding one, on the assumption that the forced vertical velocity is the observed horizontal velocity multiplied by the contour gradient. Forced vertical motion at the ground is shown by isolines; for easier interpretation of the map, areas of ascending air are shaded, and areas of descending air are left unshaded.

These charts illustrate the essence of the new mathematics: to "derive one map from another, just as one usually derives one equation from another" (Bjerknes, 1914, p. 14). As presented in *Kinematics* (Bjerknes *et al.,* 1911), the graphical calculus consisted of a great variety of procedures, many of them involving specially prepared graph paper, overlays, auxiliary charts, numerical tables, and even mechanical devices, such as the "integration machines" of J. W. Sandström.

Since it was a straightforward matter to represent the values of the seven variables on maps and thus to prepare a complete graphical description of the atmosphere at one point in time, and since Bjerknes had shown, as in the preceding illustration, that one could use the maps to find change in time, the graphical calculus gave some promise of being a suitable tool for the great task of calculating the weather. Bjerknes, it must be said, never underestimated the magnitude of that task. In 1913 he said

> Our problem is, of course, essentially that of predicting future weather. "But," says our critic, "How can this be of any use? The calculations must require a preposterously long time. Under the most favorable conditions it will take the learned gentlemen perhaps three months to calculate the weather that nature will bring about in three hours. What satisfaction is there in being able to calculate to-morrow's weather if it takes us a year to do it?"
>
> To this I can only reply: I hardly hope to advance even so far as this. I shall be more than happy if I can carry on the work so far that I am able to predict the weather from day to day after many years of calculation. If only the calculation shall agree with the facts, the scientific victory will be won. Meteorology would then have become an exact science, a true physics of the atmosphere. When that point is reached, *then* the practical results will soon develop. (Bjerknes, 1914, p. 14)

Bjerknes and others, most of them in Norway, continued through five decades to develop the graphical calculus, which they saw as potentially comparable to Newton's and Leibniz's algebraic calculus, both in power and in range of applicability. For example, the Danish meteorologist V. H. Ryd, in a book published in 1923, showed how the graphical calculus together with numerical techniques could be used to calculate air movements (Whipple, 1924a). The culmination of this development within meteorology came in the 1950s, when the graphical calculus was successfully used in forecasting (as described in Chapter 11). But it appears that Bjerknes himself never attempted the great task.

A Turn toward Practical Forecasting

Wartime conditions in Leipzig and the possibility of founding a geophysical institute in his native country prompted Bjerknes's move to Bergen in 1917. At that time he was still committed to trying to deduce the consequences of the physical laws. Having acknowledged that making forecasts in this way was a distant goal, he had undertaken "a whole series of preparatory individual problems" (Bjerknes, 1914, p. 14).

Bjerknes soon, however, changed the direction of his efforts. A particularly bad harvest in Norway in 1917 was rendered more serious by the wartime interruption of commerce. This made the Norwegian government eager to support any means of increasing agricultural production, and one way was by improving weather forecasts. Bjerknes, seeing an opportunity to provide an important service and to gain government funding for his new institute, turned from theoretical meteorology to practical forecasting. In a letter dated 4 June 1918 he wrote, "Life is fateful. Now I have suddenly become a 'practical' meteorologist. We shall try to do all we can in order to provide weather forecasts for farming" (in Jewell, 1981, p. 828).

In the following 3 years Bjerknes and his collaborators—mainly Jacob Bjerknes (his son), Tor Bergeron, and Halvor Solberg—contributed the following concepts to meteorological theory: cold front, warm front, occluded front, polar front, lines of convergence, and air mass. These collaborators, who were known as the Bergen School, and these concepts, which were referred to collectively as air-mass analysis, are an important part of the story of the expansion of meteorology that took place in the period between the world wars and which is the subject of Chapter 7. What these concepts provided was a higher-level language of the atmosphere that was not built up from physics. It is ironic that the man who became known as the advocate of calculating the weather, and as the advocate of meteorology based on the laws of physics, was also the man who initiated the development of a set of effective techniques that were neither algorithmic nor based on the laws of physics.

Bjerknes's advance on this new tack is one reason he did not attempt the great task of calculating the weather. A second reason is that he learned of Lewis Fry Richardson's attempt. In 1922 Richardson published *Weather Prediction by Numerical Process,* which described both a scheme for computing the weather and a trial forecast made using the scheme. The fact that Richardson had taken 6 weeks to compute a 6-h advance of the weather and that the resulting forecast was horribly in error convinced Bjerknes that the final goal of calculating the weather was extremely distant. But the influence was first in the other direction: Richardson, who used the same variables and essentially the same equations as Bjerknes, was convinced by Bjerknes's writings that one could predict the weather on the basis of these equations.

Chapter 6 | Lewis Fry Richardson

The First Person to Compute the Weather

In the early 20th century only a few sciences, notably astronomy, geodesy, and meteorology, involved much computation, and even with these sciences not many people thought scientific advance depended on improvement in computation. One of the first scientists to regard computation as centrally important and to envision large-scale computation was Lewis Fry Richardson. A picture he drew in 1922 of thousands of human computers working together to produce a weather forecast has become classic:

> After so much hard reasoning, may one play with a fantasy? Imagine a large hall like a theatre, except that the circles and galleries go right round through the space usually occupied by the stage. The walls of this chamber are painted to form a map of the globe. The ceiling represents the north polar regions, England is in the gallery, the tropics in the upper circle, Australia on the dress circle and the antarctic in the pit. A myriad computers are at work upon the weather of the part of the map where each sits, but each computer attends only to one equation or part of an equation. The work of each region is coordinated by an official of higher rank. Numerous little "night signs" display the instantaneous values so that neighbouring computers can read them. Each number is thus displayed in three adjacent zones so as to maintain communication to the North and South on the map. From the floor of the pit a tall pillar rises to half the height of the hall. It carries a large pulpit on its top. In this sits the man in charge of the whole theatre; he is surrounded by several assistants and messengers. One of his duties is to maintain a uniform speed of progress in all parts of the globe. In this respect he is like the conductor of an orchestra in which the instruments are slide-rules and calculating machines. But instead of waving a baton he turns a beam of rosy light upon any region that is running ahead of the rest, and a beam of blue light upon those who are behindhand. (Richardson, 1922, p. 219)

Richardson's research and vision had important effects on the development of meteorology, yet this science was his main concern for only about a dozen years.

Into and out of Meteorology

Lewis Fry Richardson was born 11 October 1881 at Newcastle upon Tyne. His father, David Richardson, was a tanner; his mother, born Catherine Fry, came from a family of grain merchants. Lewis attended Bootham School at York, where, he reported, the meteorologist J. Edmund Clark "gave us glimpses of the marvels of science" (in Gold, 1954, p. 218). After leaving Bootham in 1898, Richardson studied 2 years at Durham College of Science in Newcastle and then entered King's College, Cambridge. He did not specialize in any one science, but studied physics (under J. J. Thomson and G. F. C. Searle), chemistry, zoology, botany, and geology. He left King's in 1903 with a "First" in the Natural Science Tripos Part I.

In the next 10 years Richardson had various employments (almost all of which are named in Figure 1). Twice, each time for 1 year, he worked as a physics instructor; although conscientious, he was apparently not gifted as a teacher. He worked briefly as a mathematics assistant to Karl Pearson; it may have been from Pearson that Richardson learned statistical techniques. Twice he was an assistant at the National Physical Laboratory, first in the metallurgy division, then in the

Year	Employment	Mathematics	Meteorology	Psychology and Peace Research	Instrumentation	Other
1903	assistant in metallurgy at the NPL					
	physics instructor at					
1905	University College, Aberystwyth					L
						L
	chemist at National Peat Industries	A				A
	mathematical assistant to K. Pearson.	2A				
	assistant in meteorology at the NPL					
1910	researcher at Sunbeam Lamp Company	A				
		A				
	physics instructor at Municipal					
	School of Technology, Manchester		L	L		
			2L			
1915	Superintendent of		L		A	
	Eskdalemuir Observatory					
	member of the Friends Ambulance			L		
	Unit with the French Army		L, 2A	B	A	L
1920		L	2L, 3A			
	meteorologist at Benson Observatory	L	3A			
			L, 2A, B			
			L, A		L	
			2A, 4L			
1925		A	3A			
	lecturer in physics and mathematics		4A	L	A	
	at Westminster Training College	A	2A			
				A	2A	A
			A	2A		
1930			2A	L, 3A		
			L	L, A		L
	Principal of Paisley Technical College			2L	L, 2A	
1935						
				2L	A	
				L	3A	
				L, A		
1940				2L, A		
	in retirement at Paisley			L		
				2L		
			L			
				L		
1945				L, A		
		A		L,		
				L, A, B		
	in retirement at Kilmun		L	2A		
			L	L		
1950		2A		3A,.B		L
			L	L		
			L, A	2A		A
1953				2A		

Figure 1 This is a listing of Richardson's publications and employment as a function of time. Each of Richardson's publications was classified as concerned primarily with mathematics, with meteorology, with instrumentation, or with psychology or peace research. B denotes book, A article, and L letter.[1] Knowing the delay associated with publication, we expect a stronger correlation between employment and publications if the latter are shifted a year or so back. The small slanted lines to the right of the column of employments suggest this shift.

metrology division. And twice he worked as a researcher in industry, 2 years with National Peat Industries and 3 years with Sunbeam Lamp Company. It was in these two positions that Richardson discovered several ways to obtain approximate solutions to differential equations.

In 1909 Richardson married Dorothy Garnett, daughter of William Garnett, who had been Clerk Maxwell's demonstrator at the Cavendish Laboratory. The Richardsons had no children of their own, but adopted two boys and a girl. Dorothy became for Richardson an important support, the more needed because of his impractical nature and the fact that he often worked in intellectual isolation. In many of his works, Richardson thanks his wife for doing calculations.[2]

In 1913 Richardson became Superintendent of the Eskdalemuir Observatory, which had been built to make geomagnetic measurements but which had become a meteorological station as well.[3] Richardson, who seems to have been attracted to meteorology because he thought that one of his mathematical techniques could there be applied with great success, became fascinated with meteorology in all its aspects and came to make important contributions to meteorological theory, observation, and instrumentation.

Richardson was a Quaker and, in World War I, a conscientious objector. With some difficulty he obtained his release from the Meteorological Office and served from 1916 to 1919 as a driver with the Friends' Ambulance Unit attached to the French army. It was in the period 1913 to 1919 that Richardson worked out a scheme for computing the weather. In 1922 the scheme and a trial computation were made public in the book *Weather Prediction by Numerical Process.* It was also during the war years that Richardson first worked out a mathematical theory of human conflict, which he wrote up at book length. Richardson's account is as follows: "There was no learned society to which I dared to offer so unconventional a work (*Mathematical Psychology of War*). Therefore I had 300 copies made by multigraph, at a cost of about £ 35, and gave them nearly all away. It was little noticed" (in Gold, 1954, p. 231).

After the war Richardson worked for a short time with the Meteorological Office, but felt compelled to resign, because of his pacifistic convictions, when, in 1920, the Meteorological Office was transferred to the Air Ministry. He accepted a teaching position, which he held for 9 years, at Westminster Training College in London. There, although he worked a great deal directing students in doing experimental work and exercised his ingenuity in designing demonstrations for his physics classes, he had time to carry out his own research. Richardson was active in the Royal Meteorological Society, attending and frequently contributing to the monthly scientific meeting, serving on several committees and as honorary secretary, and refereeing papers and reviewing books. He was a member also of the Physical Society and of the Institute of Physics. In 1926 he was elected to the Royal Society.

In 1926 Richardson made a deliberate, but not quite complete, break with meteorology, and psychology became his main interest. He studied as an external student at University College, London, and in 1919 received a B.Sc. in psychology. He devised ways to quantify the intensity of mental imagery and the perception of color, loudness, touch, and pain. In 1929 Richardson became Principal of Paisley Technical College in southwest Scotland. He remained in this position until his retirement in 1940. In these years he had a heavy teaching load as well as being head administrator.

In the mid 1930s Richardson returned to the study of how wars occur.[4] Here he employed two quite different methodologies. One approach was to write differential equations to describe such things as hostility between peoples and expenditures on armaments and then to investigate what the equations predicted.[5] Richardson's other approach was to amass data on what he called "deadly quarrels," that is, all human conflicts, from murders to world wars, that result in death. He then looked for correlations between quarrels and things such as the existence of a common language or religion, the length of a common border, and the amount of money spent preparing for quarrels.

In 1940, at age 58, Richardson retired as Principal of Paisley Technical College in order to give all his time to peace research. Before his death in 1953 he completed two books, *Arms and Insecurity* and *Statistics of Deadly Quarrels,* but they were not published until 1960. The former presents mathematical theories of arms races; the latter contains an immense amount of data on human conflicts that Richardson laboriously collected over more than a decade. These books attracted notice, and in the 1960s a number of researchers took up Richardson's theoretical and statistical approaches to the study of wars.

Figure 1 provides an overview of Richardson's publications and employment. It shows how productive of published research he was in different periods of his life, suggests the range of his interests, and indicates a relation between employment and publications. Moreover, it allows one to judge the following summary of Richardson's career: first a mathematician, then a meteorologist, and finally a peace-researcher—and an improver of instruments as long as he was employed where there were laboratory instruments. However, an important part of his career, his work as educator, is underrepresented in the figure because it seldom resulted in publications.

It is interesting that Richardson, who in his daily life seemed not a practical man but a man of principles and of vision, regularly showed a concern for the practical application of his scholarly work. For Richardson the sine qua non of his scheme of weather prediction was that it actually work. He wanted his work on human conflict to be of immediate practical value, and he thought it did. In 1939 he sent a paper on arms races to a journal and told the editors that its rapid publication might help to avert the impending war; the editors rejected the paper (Platzman, 1967, p. 543).

Richardson approached the questions raised by his work in industry, the questions of meteorology, and the question of why nations go to war in the same way: he sought to express observations and ideas in a mathematical language and then use that language to draw some conclusions. In 1919 he had this to say about his approach:

> To have to translate one's verbal statement into mathematical formulae compels one to scrutinize the ideas therein expressed. Next the possession of formulae makes it much easier to deduce the consequences. In this way absurd implications, which might have passed unnoticed in a verbal statement, are brought clearly into view and stimulate one to amend the formula. Mathematical expressions have, however, their special tendencies to pervert thought: the definiteness may be spurious, existing in the equations but not in the phenomena to be described; and the brevity may be due to the omission of the more important things, simply because they cannot be mathematized. Against these faults we must constantly be on our guard (in Ashford, 1985, p. 61).

Discovery of a Numerical Method

Richardson, while employed as a scientist by industry, felt acutely the need for ways to solve differential equations. He used various analytic procedures, that is, procedures that use the techniques of mathematical analysis and yield exact algebraic solutions. However, such procedures very often could not be applied, and Richardson wrote

> There is obviously a demand for a method of solving that group of partial differential equations—of which we may regard Laplace's as the simplest type—which shall, if necessary, sacrifice accuracy above 1 per cent., to rapidity, freedom from the danger of large blunders, and applicability to more various forms of boundary surface. (1908a, p. 238)

Richardson invented a graphical procedure, yielding an approximate solution, which he described in two papers published in 1908.

Richardson's job at National Peat Industries involved deciding what drainage channels should be cut in peat beds. To aid in making this decision he devised "a simple method for determining the relation between the distance apart of ditches and the height to which the saturating water will rise with a given rainfall" (Richardson, 1908b, p. 295). Richardson performed experiments to measure the porosity of the peat and to justify the assumption that the water velocity is proportional to the pressure gradient. Then he used the principles of hydrodynamics to get a set of partial differential equations describing the flow of water. The analytic solution of the equations was attempted and "given up as hopeless." Richardson then devised a freehand graphical method, which is illustrated in Figure 2.

The draftsman, using a pencil, draws lines starting and ending at specified points on the boundary. By making several starts and by repeated amending of each, he arrives at a drawing having the following two properties: the corners of the curvilinear figures are at right angles, and the ratio of height to width is the same for all the figures. Richardson claims that one can even estimate the error involved: "The difference, then, between the selected graph and the second best

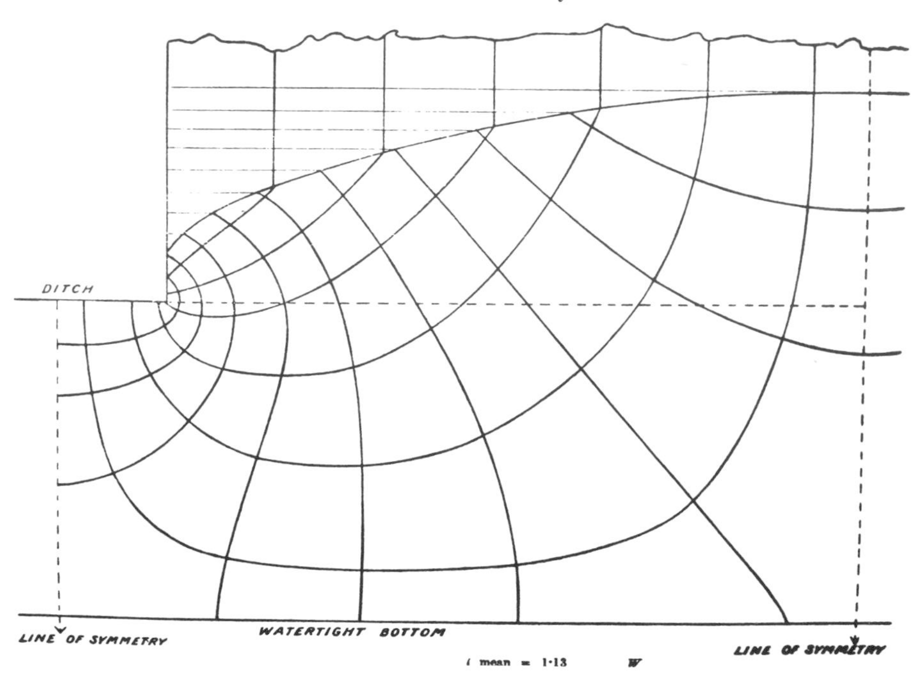

Figure 2 This drawing (from Richardson, 1908b, p. 306) illustrates a graphical method of solving a partial differential equation.

graphs is a measure of the errors of the latter and an outside limit to the errors of the former" (1908a, p. 269).

Three years later Richardson (1911) published a description of a different graphical method for solving differential equations, one requiring less skill on the part of the draftsman. In the meantime, however, he had made a discovery that would cause him to all but forget graphical procedures: in 1910 he found an arithmetical procedure that yields an approximate solution. It was this computational procedure for solving differential equations that turned Richardson toward meteorology and that became the engine for his scheme of weather prediction.

Beginning in the 1890s several mathematicians, including Carl Runge and Wilhelm Kutta, investigated the solution of ordinary differential equations by replacing derivatives with ratios of finite differences (dx/dt with $\Delta x/\Delta t$, for example). The advantage of converting differential equations to difference equations is that they can then be solved by arithmetical procedures; the disadvantage is that the solution is only approximate. What Richardson did was to develop a similar procedure for partial differential equations. He first described the arithmetical procedure in a 1910 paper entitled "The approximate arithmetical solution by finite differences of physical problems involving differential equations, with an application to the stresses in a masonry dam." Figure 3 shows some of the results presented in this paper.

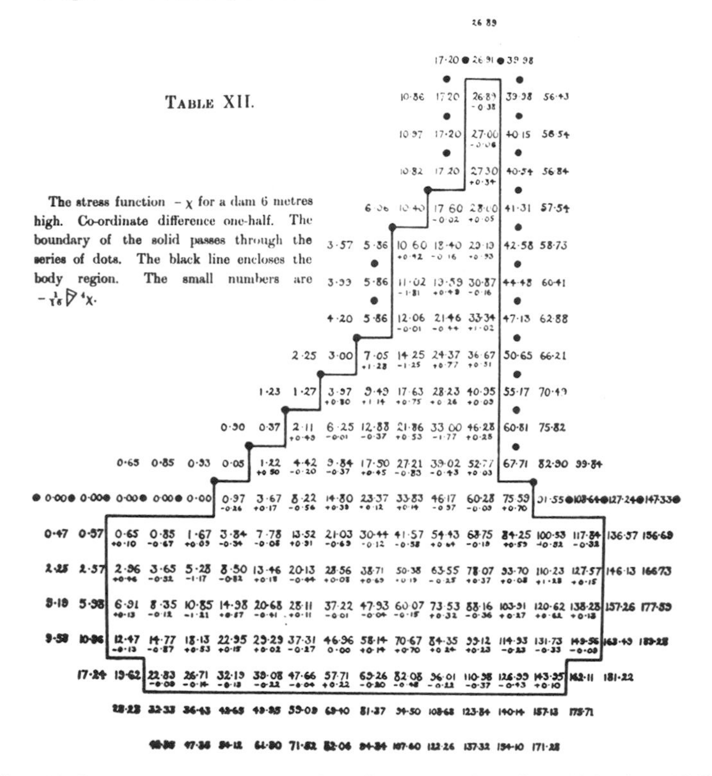

Figure 3 This figure represents a cross section of a masonry dam (from Richardson, 1910, p. 343). The water is to the right; the surface of the earth to the left of the dam is indicated by the row of zeroes. The numbers at each position are predictions of the stresses there.

Notice Richardson's method: partition space into rectilinear cells, assume the variables are constant within a cell, and solve the differential equations by replacing derivatives by ratios of finite differences. Several years later Richardson applied exactly this method to predicting the weather.

It is clear in these early papers that Richardson is addressing scientists in industry. For example, he writes, "So far I have paid piece rates for the operation $\partial_x^2 + \partial_x^2$ of about n/18 pence per co-ordinate point, n being the number of digits. The chief trouble to the computers has been the intermixture of plus and minus signs.[6] As to the rate of working, one of the quickest boys averaged 2,000 operations $\partial_x^2 + \partial_x^2$ per week, for numbers of three digits, those done wrong being discounted" (Richardson, 1910, p. 325).

It is interesting to compare the arguments Richardson had earlier made for the graphical method with those he made for the arithmetical method. About the graphical method he wrote

> Further than this, the method of solution must be easier to become skilled in than are the usual methods with harmonic functions. Few have time to spend in learning their mysteries. And the results must be easy to verify—much easier than is the case

> with a complicated piece of algebra. Moreover, the time required to arrive at the desired result by analytical methods cannot be foreseen with any certainty. It may come out in a morning, it may be unfinished at the end of a month. (1908a, p. 238)

The arithmetical method shares these advantages over the analytic, and has, Richardson wrote, additional advantages: whereas the graphical method requires some skill in draftsmanship, the arithmetical method can be carried out by people unskilled in draftsmanship as well as algebra; with the arithmetical method there is no lower bound on the error, whereas with the graphical the error can hardly be brought under one percent.

A Scheme to Compute the Weather

Richardson had a broad training in the sciences and had already worked professionally in a variety of areas. In 1911 he began to think about meteorology as an area in which his method of solution might be fruitful. In the preface to *Weather Prediction by Numerical Process,* he says as much: "This investigation grew out of a study of finite differences. . . ." It seems that Richardson, having made a tool, looked for something to do with it.

Meteorology has long been an important breeding ground for mathematical and physical ideas; statistical methods, numerical methods, and ideas about turbulence have come out of meteorology. At the same time meteorology has long exercised a great attraction on scientists with new mathematical and physical ideas. We saw in Chapter 5 that Bjerknes was attracted to meteorology because he believed his circulation theorems could there be applied fruitfully. Seventy years earlier James Espy, after studying the effect of heat on the expansion of air, turned to meteorology because he believed he had found the "lever with which meteorology was to move the world" (in Kutzbach, 1979, p. 22). In the 1860s three scientists, William Thomson, Theodore Reye, and H. Peslin, each of whom was investigating the implications of the first law of thermodynamics, turned to meteorology with the conviction "that here an avenue had opened which would lead to significant advancement of meteorology" (Kutzbach, 1979, p. 46). And just as Richardson was attracted to meteorology as an area in which to apply his numerical methods, so 30 years later John von Neumann was similarly moved, as we will see in Chapter 10, because of his desire to demonstrate the usefulness of the electronic computer.

Richardson's decision to turn to meteorology was influenced by Vilhelm Bjerknes. Although most meteorological observations were, from the late 17th century on, *measurements,* meteorological theory became quantitative only gradually, mainly as a result of applying mathematical physics to atmospheric phenomena. This process hardly began before 1850, yet, as we saw in Chapter 5, by the early 20th century Bjerknes was arguing vociferously that the weather could be *calculated* using the physical laws. Bjerknes thought that observational meteorology had reached its goal: it was able to give a complete characterization of the atmosphere at a given place and time. This it could do by measuring seven quantities:

the three components of the velocity of air v_E, v_N, and v_H, the pressure p, the temperature θ, the mass density ρ, and the water content per unit volume μ.

Richardson demurred, although only briefly. A complete account, he argued, requires an eighth variable: the amount of dust in the air (Richardson, 1922, p. 59). Whether clouds form or rain falls is partially dependent on the concentration of dust in suspension, and the effect of dust on meteorological phenomena was already an established subject of research. In the last decades of the 19th century the English scientist John Aitken did a great deal of research in this area. But the difficulty of measuring this variable and the difficulty of relating it quantitatively to cloud formation and rainfall led Richardson to leave it out of his scheme for predicting the weather.

Richardson, following Bjerknes, set out to formulate seven equations that would completely determine the behavior of the atmosphere given its initial state. The seven equations involve the seven dependent variables, the four independent variables (three variables specifying position and one variable specifying time), and known constants. Figure 4 gives Richardson's set of equations.

eastward dynamical equation

$$-\frac{\partial m_E}{\partial t} = \frac{\partial p}{\partial e} + \frac{\partial}{\partial e}(m_E v_E) + \frac{\partial}{\partial n}(m_E v_N) + \frac{\partial}{\partial h}(m_E v_H) - 2\omega \sin\phi\, m_N + 2\omega \cos\phi\, m_H + \frac{3 m_E v_H}{a} - \frac{2 m_E v_N \tan\phi}{a}$$

northward dynamical equation

$$-\frac{\partial m_N}{\partial t} = -g_N \rho + \frac{\partial p}{\partial n} + \frac{\partial}{\partial e}(m_N v_E) + \frac{\partial}{\partial n}(m_N v_N) + \frac{\partial}{\partial h}(m_N v_H) + 2\omega \sin\phi\, m_E + 2\omega \cos\phi\, m_H + \frac{3 m_N v_H}{a} - \frac{\tan\phi\,(m_E v_E - m_N v_N)}{a}$$

upward dynamical equation

$$-\frac{\partial m_H}{\partial t} = g\rho + \frac{\partial p}{\partial h} + \frac{\partial}{\partial e}(m_H v_E) + \frac{\partial}{\partial n}(m_H v_N) + \frac{\partial}{\partial h}(m_H v_H) - 2\omega \cos\phi\, m_E + \frac{2 m_H v_H - m_E v_E - m_N v_N - m_H v_N \tan\phi}{a}$$

conservation-of-mass equation

$$-\frac{\partial \rho}{\partial t} = \frac{\partial m_E}{\partial e} + \frac{\partial m_N}{\partial n} - \frac{m_N \tan\phi}{a} + \frac{\partial m_H}{\partial h} + \frac{2 m_H}{a}$$

conveyance-of-water equation

$$\rho \frac{d\mu}{dt} = \frac{\partial w}{\partial t} + \frac{\partial}{\partial e}(w v_E) + \frac{\partial}{\partial n}(w v_N) - \frac{w v_N \tan\phi}{a} + \frac{\partial}{\partial h}(w v_H) + \frac{2 w v_H}{a}$$

conveyance-of-heat equation

$$\frac{dq}{dt} = b\frac{d\theta}{dt} + p\frac{d}{dt}(1/\rho)$$

characteristic gas equation

$$p = (c + k\mu)\rho\theta$$

Figure 4 These are the main equations in Richardson's scheme of weather prediction. e, n, h, and t (representing, respectively, eastward distance, northward distance, height, and time) are the independent variables. By definition, $m_E = \rho v_E$, $m_N = \rho v_N$, $m_H = \rho v_H$, and $w = \mu\rho$. ϕ is the latitude, and a, b, c, and k are known constants. dq/dt is specified as a sum of terms, each one involving only the other variables and known constants. Hence the formulation is mathematically complete, there being seven equations in the seven unknowns v_E, v_N, v_H, ρ, p, θ, and μ.

The first three equations are essentially Newton's "force equals mass times acceleration." The fourth equation says that if the mass-density decreases at a place, then matter must have moved away; the fifth says the same for the water content. The sixth says that an addition of heat must either raise the temperature or do work or both. The seventh is a combination of Boyle's law (that pressure is inversely proportional to volume) and Charles' law (that volume is directly proportional to absolute temperature), with an allowance made for the presence of water vapor.

Richardson acknowledged his debt to Bjerknes, "The extensive researches of V. Bjerknes and his School are pervaded by the idea of using the differential equations for all that they are worth. I read his volumes on *Statics and Kinematics** soon after beginning the present study, and they have exercised a considerable influence throughout it. . . ." (Richardson, 1922, p. xii).[7] Recall that Bjerknes, who believed that an analytic solution of the equations was out of the question, thought graphical procedures would eventually suffice. We will see below what lessons Bjerknes drew from Richardson's work.

It is noteworthy that the belief, which came to be held by many meteorologists, that this system of equations fully characterizes the action of the atmosphere did not engender a sort of Scholasticism, a turning of one's back on observational evidence. Part of the reason for this was the mathematical intractability of the equations. Bjerknes, Richardson, and others had to make repeated simplifications to get answers from the equations, and the acceptability of each simplification was judged by reference to observational evidence.

The Data Requirements

Even if the equations are valid and complete, they cannot be used to predict the weather without an accurate and complete description of the atmosphere at a particular moment. Richardson found that the available data were much too sparse for his purposes. One may wonder at this, knowing that scientists interested in the weather had long had access to exceptionally large amounts of observational data and that many meteorologists had given much of their time, often in an exceptionally faithful and painstaking fashion, to the gathering of more. Nevertheless it seems that whenever anyone had a plan for making use of the existing data, he found he needed still more. This was especially true of Richardson.

Richardson wanted more data, not to build an archive of past weather nor to discover empirically some regularity, but simply because his computational scheme required it. On a map of Europe he drew a checkerboard pattern, which he regarded as partitioning the atmosphere horizontally. He further divided the atmosphere vertically into five layers, so that each square was divided into five blocks. He regarded the atmosphere in each block as represented by a single value of pressure, a single value of temperature, and so on. This "smoothing" of the data meant that atmospheric phenomena—such as thunderstorms—of extent less than the extent of a block could not be represented.

In Richardson's scheme the derivatives in the differential equations are replaced by ratios of finite differences. For the spatial coordinates the finite differences are given by the division into blocks; for the time coordinate Richardson took Δt to be 6 h. The scheme requires values of certain of the seven variables in each of the blocks at time t. It gives the values of the same variables at time $t + \Delta t$. Here was a grave problem: there were no weather stations in many of the squares, and there were few data for the four layers above the surface layer. Richardson called attention to the contrast between the two maps in Figure 5.

Richardson gave much attention to reducing the error caused by replacing derivatives with finite differences. He regularly used a method of "centered differences" that required that different variables be evaluated at different points. For this reason he specifies that three of the seven dependent variables (the components of the velocity of the air) are to be measured at the center of each white square (or, for Figure 5B, each square marked "M"), and the remaining four at the center of each shaded square (or each square marked "P").

In order to test his scheme Richardson needed two complete sets of data, separated in time by 6 h. The completeness of the data was essential: "For a purpose such as numerical prediction by finite differences, meteorological observations are useless if they are not very complete" (Richardson, 1923, p. 345). Richardson was fortunate in having unusually complete data for 20 May 1910, since this had been an "international balloon day" when a number of European meteorological stations gathered upper-air data at specified times. Even so, he had to do a good deal of interpolating and extrapolating to fill out the table of initial values shown in Figure 6.[8]

Richardson did much work to improve observational techniques, and it is clear that much of this was motivated by the needs of his computational scheme. For example, he invented an instrument, which he called a contrast photometer, to measure optically the water content of clouds. The article reporting the new technique begins as follows: "Since the initial data on which a weather prediction is to be based must, to be adequate, include the water-content of the clouds; and since observations of this are scanty, it will be well to survey what can be done in this respect" (Richardson, 1919).

Another example is provided by a technique Richardson invented for measuring wind speeds high in the atmosphere. Again, it is clear he undertook this work because his scheme required the data (see Richardson, 1923, p. 345).[9] The actual observations are fairly easy: spherical bullets of different diameters are shot upward, and the gun is tilted until the bullets fall back to the gun.[10] Using the measured angles-of-tilt for bullets of different sizes and for different amounts of gun-

Figure 5 (A) Richardson's proposed arrangement of meteorological stations; if suitability to his scheme were the only criterion for placement of observing stations, then there would be one station in each square, that station at the square's center. (B) The actual arrangement of meteorological stations at the time of Richardson's work. (These maps are from Richardson, 1922, frontispiece and p. 184.)

A

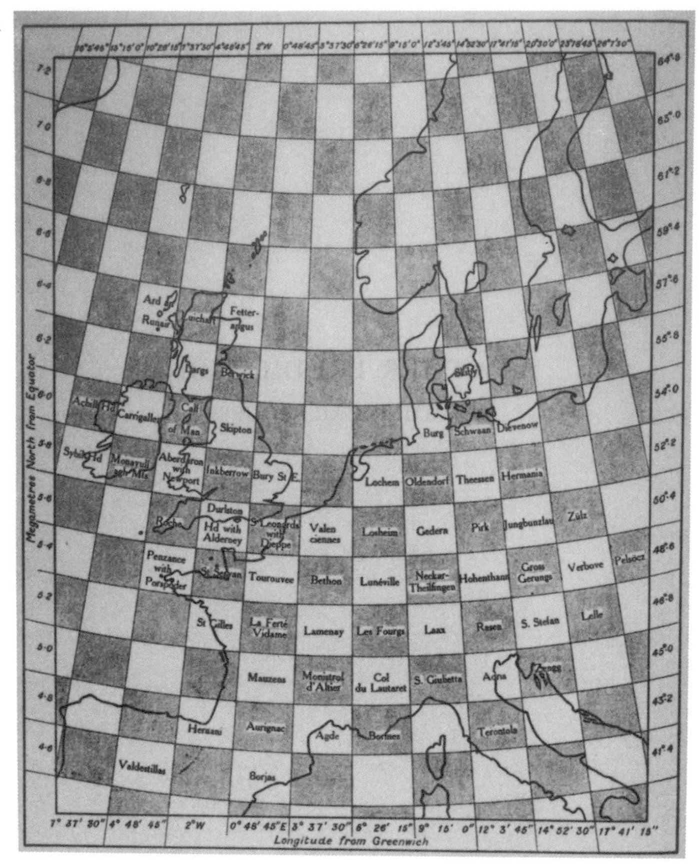

Megametres North from Equator
Longitude from Greenwich
7° 31' 30"
4° 48' 45"
2°W
0° 48' 45"E
3° 37' 30"
6° 26' 15"
9° 15' 0"
12° 3' 45"
14° 52' 30"
17° 41' 15"
7·2
7·0
6·8
6·6
6·4
6·2
6·0
5·8
5·6
5·4
5·2
5·0
4·8
4·6
64°·6
63°·0
61°·2
59°·4
57°·6
55°·8
54°·0
52°·2
50°·4
48°·6
46°·8
45°·0
43°·2
41°·4
Ard
Runan
Laichart
Fetter-angus
Dargs
Berwick
Achill Hd
Carrigallen
Calf of Man
Skipton
Burg
Schwaan
Dievenow
Sybil Hd
Monavullagh Mts
Aberdaron with Newport
Inkberrow
Bury St E.
Lochem
Oldendorf
Theessen
Hermania
Roche
Durlston Hd with Alderney
St Leonards with Dieppe
Valenciennes
Losheim
Gedern
Pirk
Jungbunzlau
Zülz
Penzance with Porspoder
St Servan
Tourouvre
Bethon
Lunéville
Neckar-Theilfingen
Hohenthann
Gross Gerungs
Verbove
St Gilles
La Ferté Vidame
Lamenay
Les Fourgs
Laax
Rasen
S. Stefan
Lelle
Mauzens
Monistrol d'Allier
Col du Lautaret
S. Giubetta
Herauni
Aurignac
Agde
Teroutola
Valdestillas
Borjas

B

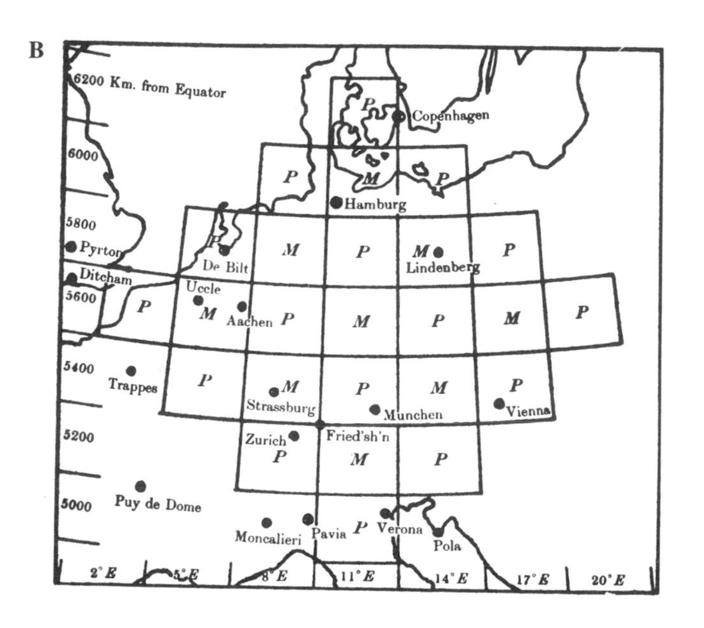

6200 Km. from Equator
6000
5800
5600
5400
5200
5000
Copenhagen
Hamburg
Pyrton
De Bilt
Lindenberg
Ditcham
Uccle
Aachen
Trappes
Strassburg
Munchen
Vienna
Zurich
Fried'sh'n
Puy de Dome
Moncalieri
Pavia
Verona
Pola
P
M
2° E
5° E
8° E
11° E
14° E
17° E
20° E

CH. 9/1 TABLE OF INITIAL DISTRIBUTION 185

Obtained by interpolating or extrapolating from observations taken at 1910 May 20ᵈ 7ʰ G.M.T.

kilometres north from equator	longitude 5° E	longitude 8° E	longitude 11° E	longitude 14° E	longitude 17° E
6000		θ_1 214° A	1000× 1000× M_{E20} -65 M_{N20} +8 M_{E42} +127 M_{N42} -104 M_{E64} +81 M_{N64} -25 M_{E86} -81 M_{N86} zero M_{EG8} -198 M_{NG8} +84 h_G = 0	θ_1 216° A	NOTE: The following stratosphere temperatures have also been used long 11° E, lat 6200 N, 216° ,, 9° E, ,, 5600 N, 217° ,, 20° E, ,, 5600 N, 216°
5800	θ_1 215° A	1000× M_{E20} -70 M_{E42} -62 M_{E64} -114 M_{E86} -91 M_{EG8} -160 h_G 15000	100× θ_1 214 A p_2 2047 W'_{42} 0·0 p_4 4090 W'_{64} 0·2 p_6 6086 W'_{86} 0·4 p_8 7983 W'_{G8} 0·9 p_G 9883 h_G 20000	1000× M_{N20} -160 M_{N42} +40 M_{N64} -60 M_{N86} -60 M_{NG8} -219 h_G 10000	θ_1 216° A
5600	1000× 1000× M_{E20} -30 M_{N20} -110 M_{E42} -245 M_{N42} +300 M_{E64} -223 M_{N64} +158 M_{E86} -91 M_{N86} +87 M_{EG8} -18 M_{NG8} +15 h_G 20000	100× θ_1 212° A p_2 2047 W'_{42} 0·0 p_4 4083 W'_{64} 0·2 p_6 6067 W'_{86} 0·5 p_8 7950 W'_{G8} 1·2 p_G 9834 h_G 20000	1000× 1000× M_{E20} -56 M_{N20} -18 M_{E42} -146 M_{N42} 62 M_{E64} -95 M_{N64} +29 M_{E86} 52 M_{N86} +58 M_{EG8} -110 M_{NG8} +55 h_G 40000	100× θ_1 214° A p_2 2049 W'_{42} 0·0 p_4 4091 W'_{64} 0·1 p_6 6087 W'_{86} 0·4 p_8 7979 W'_{G8} 0·9 p_G 9768 h_G 30000	1000× M_{E20} -100 M_{N20} -32 M_{E42} zero M_{N42} -260 M_{E64} -55 M_{N64} -135 M_{E86} -25 M_{N86} +48 M_{EG8} -190 M_{NG8} +160 h_G 30000
5400	100× θ_1 214° A p_2 2030 W'_{42} 0·0 p_4 4049 W'_{64} 0·2 p_6 6044 W'_{86} 0·7 p_8 7928 W'_{G8} 1·4 p_G 9744 h_G 20000	1000× M_{E20} +27 M_{E42} -326 M_{E64} 186 M_{E86} -33 M_{EG8} +48 h_G 40000	100× θ_1 212° A p_2 2050 W'_{42} 0·0 p_4 4090 W'_{64} 0·1 p_6 6079 W'_{86} 0·4 p_8 7960 W'_{G8} 0·9 p_G 9626 h_G 40000	1000× M_{N20} zero M_{N42} -166 M_{N64} -95 M_{N86} -19 M_{NG8} 65 h_G 40000	100× θ_1 214° A p_2 2044 W'_{42} zero p_4 4082 W'_{64} 0·1 p_6 6068 W'_{86} 0·4 p_8 7978 W'_{G8} 0·9 p_G 9682 h_G 20000
5200		100× θ_1 214° A p_2 2039 W'_{42} 0·0 p_4 4062 W'_{64} 0·2 p_6 6050 W'_{86} 0·6 p_8 7949 W'_{G8} 0·9 p_G 8746 h_G 120000	1000× 1000× M_{E20} -50 M_{N20} +80 M_{E42} -280 M_{N42} +41 M_{E64} -175 M_{N64} +150 M_{E86} -105 M_{N86} +80 M_{EG8} -155 M_{NG8} +40 h_G 180000	100× θ_1 214° A p_2 2043 W'_{42} 0·0 p_4 4075 W'_{64} 0·1 p_6 6065 W'_{86} 0·4 p_8 7967 W'_{G8} 1·0 p_G 8458 h_G 150000	
5000 km north of equator			100× θ_1 218° A p_2 2034 W'_{42} 0·0 p_4 4034 W'_{64} 0·2 p_6 6031 W'_{86} 0·6 p_8 7957 W'_{G8} 1·4 p_G 9972 h_G 10000		

All the quantities in any one square refer to the same latitude and longitude, namely to those stated in the margin. The notation is defined in Ch. 12. A numerical value, when multiplied by the integral power of ten, if any, standing at the head of its column, is then expressed in C.G.S. units. To save space the equality symbols are omitted, thus $\overset{1000\times}{M_{E20} - 65}$ stands for $M_{E20} = -65000$ grm cm^{-1} sec^{-1}.

Figure 6 This table (from Richardson, 1922, p. 185) shows the initial values that Richardson used to test his scheme. The squares correspond to the squares of the map in Figure 5B. θ is the temperature, h is the height above mean sea level, and W is water substance per area of stratum. For each of the five strata, values are given of the pressure, denoted p, and the eastward and northward components of momentum per unit volume, denoted respectively M_E and M_N.

powder, Richardson was able to deduce the velocity of the wind at different heights. But to do so he needed to take into account other variables, such as temperature and moisture, and to distinguish their effects from the effect of the wind.[11]

Another effect of the computational scheme was that Richardson's attention was drawn to the accuracy of the data. As a result, on several occasions Richardson called for improvements in observational techniques that had been considered satisfactory. On page 128 of Richardson (1922), for example, he calculates some horizontal temperature gradients and laments that the probable errors in temperatures obtained by sounding balloons are so large that his calculations cannot be tested.

The Theoretical Basis

New ways of gathering data were important. But for some variables that could not be measured directly there was no theory at hand to allow indirect measurement; the vertical speed of the air was a notable example. And for some important atmospheric phenomena there were no quantitative theories that would allow the phenomena to be taken into account in a computation; turbulence was a notable example.

One of the variables in Richardson's scheme is v_H, the vertical component of the air's velocity. Some attempts had been made to measure the vertical speed: J. S. Dines used a pair of theodolites to measure the speed of ascent of a balloon, first in free air, then in a closed shed; he subtracted the latter speed from the former to get v_H. But such measurements were seldom made, and Richardson needed to know v_H at many geographic locations and at five different heights at each. He concluded, "If progress is to be possible it can only be by eliminating the vertical velocity" (Richardson, 1922, p. 115).

Richardson regarded the few existing theories of the vertical speed as inappropriate:

> There are various theories which arrive at the vertical velocity by treating the air as if it were not shearing, or which either neglect, or else fail to eliminate, some of the time changes at fixed points. For such reasons the otherwise interesting discussions by W. H. Dines and by M. Berek will not serve for numerical prediction. (1922, p. 115)

An entire chapter of *Weather Prediction by Numerical Process* is given to the derivation of an equation for v_H. With such an equation, one can compute v_H from the observed values of the other variables, or—what amounts to the same thing as far as the predictions are concerned and the course Richardson adopted—one can use the new equation to eliminate v_H from the set of equations. The derivation of this equation for the vertical speed was seen by some meteorologists as the greatest contribution of this book to theoretical meteorology. It is clear that Richardson undertook this work because his scheme required it.

In this case Richardson devised a theory to measure indirectly a quantity that could not be measured directly. In other cases Richardson devised a theory to provide a quantitative treatment—the only kind of treatment that could be worked into his scheme—for a phenomenon that had not previously been treated quantitatively and that had to be taken into account. His theory of turbulence is an example.

Table I

History of Richardson's Verse

Swift[a] 1733	DeMorgan[b] 1863(?)	Richardson/Ashford[c] 1922/1985
The Vermin only teaze and pinch Their foes superior by an Inch. So, Nat'ralists observe, a Flea Hath smaller Fleas that on him prey And those have smaller Fleas to bite 'em And so proceed *ad infinitum.* Thus every Poet in his kind Is bit by him that comes behind; Who, tho' too little to be seen, Can teaze, and gall, and give the Spleen.	Great fleas have little fleas upon their backs to bite 'em And little fleas have lesser fleas, and so *ad infinitum* And the great fleas themselves, in turn, have greater fleas to go on; While these have greater still, and greater still, and so on.	Big whirls have little whirls that feed on their velocity, And little whirls have lesser whirls and so on to viscosity — in the molecular sense. And the big whirls of bigger ones partake in the rotation, Until at last we reach the gen'ral circulation — in the global sense.

[a] Johnathan Swift's lines were published in his *On Poetry a Rhapsody.*

[b] Augustus DeMorgan's verse first appeared in a review of a book published in 1863.

[c] Richardson's lines (the first stanza) are from *Weather Prediction by Numerical Process,* and Ashford's are from *Prophet or Professor?*

A fluid may move in a regular fashion, such as the laminar flow of water in a narrow tube, or in an irregular fashion. The latter, which is called turbulence, is difficult to treat mathematically because the irregular movements are not restricted to a scale small enough to permit a statistical treatment, nor are they so few that they can be treated individually.

Richardson had to take turbulence into account: the momentum of wind is dissipated by turbulence, a great deal of heat is transferred by turbulence, and so on. One of the most striking features—once someone has pointed it out—of turbulence is its scale invariance: it looks pretty much the same at various magnifications. When one looks closely at an eddy, one sees smaller eddies. As Richardson put it: "Big whirls have little whirls that feed on their velocity, and little whirls have lesser whirls and so on to viscosity—in the molecular sense." [12] Since this verse is the best-known sentence from *Weather Prediction by Numerical Process,* and since it expresses a crucial insight into turbulence, perhaps we may take a moment to look at the history of the verse (see Table I).

Jonathan Swift noticed a psychological scale-invariance—every poet is criticized by slightly inferior poets—and he compared it to a biological scale-invariance. Augustus DeMorgan, who was making an analogy to a theory of the constitution of the universe, expressed the biological scale-invariance twice: fleas have smaller fleas and so on, and fleas have larger fleas and so on. Richardson and Ashford expressed a meteorological scale-invariance, and—what is new—they identified the end points of the scale. It is interesting that the flow of energy is up the scale in DeMorgan's verse, down the scale in Richardson's.[13] Indeed, this strange property of scale-invariance that Richardson noted has recently been used by scientists to put a handle on turbulence, as is discussed in Chapter 13.

Turbulence had been studied under certain conditions by the physicist Osborne Reynolds. He found a dimensionless quantity,[14] now called Reynolds' number, that provides a test for whether flow will be laminar or turbulent. Reynolds's work, however, had little immediate relevance to meteorology.

Richardson was not the first meteorologist to try to quantify turbulence in the atmosphere. G. I. Taylor, largely on the basis of empirical studies, devised an equation for the rates at which heat and moisture are transferred from one atmospheric layer to another by turbulence. But because the atmospheric layers in Richardson's scheme were so thick, Taylor's assumption of constant density within a layer was unacceptable. Thus his equation could not be used. Richardson worked long to devise an appropriate theory. One of his results was that there is a "Reynolds' number" (that is, a dimensionless quantity that provides a test for turbulence) for the atmosphere; this number is now called Richardson's number.[15]

Thus a scheme like Richardson's, which encompasses many phenomena, directs one's attention to the areas of least understanding. Other effects of such a scheme are that one is forced to simplify (when not doing so blocks the computation), to make explicit one's assumptions, and to change them when they lead one astray.[16] Richardson (1922) did frequently have to simplify. The principle Richardson states on page 156:

> In writing the chapter on the fundamental equations the ideal was to obtain a description of atmospheric phenomena which should be in the first place correct, and which, secondly, might be used in prediction. Here in Chapter 8 the order of emphasis is reversed. The ideal is now to make a scheme first workable and secondly as exact as circumstances permit.

But before Chapter 8—indeed, throughout the book—Richardson (1922) has an eye to the practical solvability of the equations. On page 118, for example, he writes: "As [Equation] (12) is inconveniently complicated, let us simplify it by making the following two approximations which will probably not affect the accuracy by one per cent."

A Test of the Scheme

Even with the necessary data and the necessary theory, Richardson had no easy task: to describe a computation, that is, a sequence of arithmetical operations, that would transform the data, in accordance with the theory, into a prediction. This task Richardson finally achieved: he found a procedure for computing a complete description of the atmosphere at time $t + \Delta t$ given a complete description at time t. Since the procedure could be applied repeatedly (and for Δt positive or negative), meteorology could then achieve what astronomy had achieved: the ability to compute a complete description for any point in time without the need to make new observations.

Roughly speaking, Richardson's scheme works as follows. Using the data describing the atmosphere at time t and the equations of Figure 4, one computes the rate of change dx/dt of variable x.[17] One then assumes that dx/dt is constant over the time interval Δt, so takes the new value of x to be $x + (dx/dt)\Delta t$.

It is, of course, not so simple. In the first place, the equations in Figure 4 are only the basic equations. Richardson added a great many terms to these equations in order to include secondary effects, some of which are mentioned below. Sometimes the secondary effects could not easily be expressed in a formula, so Richardson specified that a table be used in doing the computation.[18] In the second place, one sometimes needs to know, in computing the rate of change of a variable, not only the values of the other variables, but also some of their rates of change, as in the fifth and sixth equations in Figure 4. In the third place, and even more troublesome, there are difficulties associated with the method of centered differences that Richardson used to reduce error.

Dealing with all such difficulties—and they had to be dealt with as a whole since changes in one part of the scheme affected other parts—and finally getting an algorithm took Richardson 3 or 4 years. The magnitude of this labor—making the scheme *work*—prompted Napier Shaw to say that "when contemplating Richardson's efforts he had been reminded of the lines from Juvenal 'it is pleasant to stand on the heights and watch a ship toiling in the waves' " (in Ashford, 1985, p. 109).[19]

It is not then surprising that Richardson, when he got his scheme to work, compared it to a Heath Robinson device (quoted in Platzman, 1967, p. 546).[20] He described the process as follows:

> Imagine that you receive the parts of a machine. There are wheels, levers, casings and a hundred and one pieces that you do not recognise. Many of them are beautifully made and finished. A few are rough castings. You believe that they fit together into a machine, but you do not know what shape the machine is and the parts are not labelled. These parts symbolize the existing pieces of meteorological theory, as found in the literature. . . . you begin trying what will fit onto what and so you build up a machine. Then comes the question of fuel to drive it—the observations. Then you try to get it to go round—and that reminds me of starting a motor car at -15° C. You turn and turn the handle until your back aches and nothing happens. Then on a renewed effort the engine makes a noise like "houi" and turns once around. "Continuez" says the Frenchman, "elle a dit *oui.*" That is the stage we have got to, the wheels have gone once round, and that prompts us to continue. (Platzman, 1967, p. 546)

All the reasoning and all the data in the first 180 pages of *Weather Prediction by Numerical Process* result in a set of 23 "computing forms." Each of these specifies a sequence of arithmetical operations, and all the forms fit together to form the algorithm for predicting the weather. As Richardson put it, "The computing forms which are used for this purpose may be regarded as embodying the process and thereby summarizing the whole book." (1922, p. 181). Figure 7 is an example of a computing form.

The numbers in italics are those that are written in during a computation. The

Computing Form P I. Pressure, Temperature, Density, Water and Continuous Cloud

Longitude = *11° East* Latitude = *5400 km North* Instant *1910 May 20ᵈ 7ʰ G.M.T.*

Bar.:—		p. *185*	*Helmert*	previous		Ch. 8/2/2		p. *185*		ρ and p					
Height h	δh	p	g	R	$\rho=\frac{R}{\delta h}$	P	$p=\frac{P}{\delta h}$	W	$\mu=\frac{W}{R}$	θ	Density of saturated vapour w_s	$W_s=w_s\delta h$	Precipitated = $W-W_s-0{\cdot}4$	Q in continuous cloud	Potential temperature at surface pressure
$10^5\times$	$10^5\times$	$10^2\times$			$10^{-3}\times$	$10^5\times$	$10^5\times$		$10^{-3}\times$	°A	$10^{-6}\times$				
h_0		*0*													
			9751	*210·3*		*1278*		zero	zero	*212*					By Ch. 8/2/6 # 13
$h_2 = 11{\cdot}8$		*2050*													
	4·6		*978·2*	*209·0*	*0·454*	*1362*	*2965*	*0·0*	*0·0*	*227·5*		*0·0*	*0·0*	*0·0*	*320·8*
$h_4 = 7{\cdot}2$		*4090*													
	3·0		*979·2*	*203·3*	*0·677*	*1501*	*5003*	*0·1*	*0·5*	*257·7*	*1·3*	*0·4*	*0·0*	*0·0*	*312·0*
$h_6 = 4{\cdot}2$		*6079*													
	2·2		*980·0*	*192·1*	*0·875*	*1538*	*6990*	*0·4*	*2·1*	*279·0*	*6·0*	*1·6*	*0·0*	*0·0*	*305·8*
$h_8 = 2{\cdot}0$		*7960*													
	1·6		*980·6*	*169·9*	*1·061*	**1402*	*8770*	*0·9*	*5·3*	*287·5*	*12·3*	*2·0*	*0·0*	*0·0*	*295·3*
$h_e = 0{\cdot}4$		*9626*													

* But see Ch. 4/4 # 14.

Figure 7 This is the first of Richardson's 23 computing forms.

third column of numbers comes from the initial data [printed on page 185 in Richardson's book (1992), reproduced as Figure 6 above]. The numbers in the next two columns come from tables. The numbers in the fifth column result from dividing each entry in the fourth column by the corresponding entry in the second column. And so it goes. Some of the numbers computed here are entered onto other computing forms.

Richardson took 6 weeks to compute the change in the atmospheric variables, at only two of the squares, from 4 a.m. 20 May 1910 to 10 a.m. that same day. The results were disappointing. One predicted value, that for the change in pressure at the surface, was drastically in error: Richardson's computation led to a predicted change of 145 millibars while the actual pressure change was less than 1 millibar and the greatest range of observed pressures is just over 100 millibars. As Napier Shaw put it, "the wildest guess, therefore, at the change in this particular element would not have been wider of the mark than the laborious calculation of six weeks" (1922, p. 764). It is ironic that this forecast, which is probably the forecast that consumed the greatest amount of labor by a single person, is no doubt one of the least accurate forecasts ever.

The Necessity of Numerical Analysis

Work on *Weather Prediction by Numerical Process* included a great deal of what is now called numerical analysis. Richardson, whose entire approach depended upon his numerical method of solving partial differential equations, felt the need to argue for the respectability of this new sort of mathematics. In 1925 in *The Mathematical Gazette* he wrote

> It is said that in a certain grassy part of the world a man will walk a mile to catch a horse, whereon to ride a quarter of a mile to pay an afternoon call. Similarly, it is not quite respectable to arrive at a mathematical destination, under the gaze of a learned society, at the mere footpace of arithmetic. Even at the expense of considerable time and effort, one should be mounted on the swift steed of symbolic analysis. (p. 415)

Mathematicians had several reasons for disdaining Richardson's approach. First, they felt that their job was to discover the truth, not approximations. Second, the main business of academic mathematicians was to prove theorems, and when differential equations are attacked analytically, there is much theorem-proving to do. Mathematicians can, of course, prove theorems about methods of approximations, but this is a different style of mathematics, one that did not become common until about 1950. Third, many mathematicians must have been proud of their armamentarium of theorems that allowed them to solve many differential equations, while an arithmetical procedure that can be applied to differential equations of all sorts makes most of their weapons unnecessary—as Richardson puts it, "the bulk of the work can be done by clerks who need not understand algebra or calculus" (1910, p. 325).

In the decades since World War II the new style of mathematics has become a recognized branch of mathematics. Its main concern is finding ways to solve problems numerically. The development of methods of interpolation, the analysis of error, the study of stability of algorithms, and the comparison of algorithms are parts of numerical analysis. A study of the relation between this branch of mathematics and the technics of computing needs to be done.[21]

Several of Richardson's earliest published papers are essentially studies of graphical or numerical algorithms. Richardson's comparison of graphical methods and arithmetical methods was discussed above. Richardson made a practice of testing an algorithm by applying it to an equation having an analytic solution and comparing the exact solution with the solution given by the algorithm. He also worked to find ways of estimating the errors when the actual errors could not be found (because an exact solution could not be obtained).

At times Richardson, acting like a mathematician, proved theorems about an algorithm—concerning, for example, a bound on the error or under what conditions the algorithm can be applied. There are several examples in his 1910 paper on stresses in a masonry dam. More often Richardson behaved like an engineer: if a few tests of an algorithm came out well, he put it to immediate use and moved on to other matters.

For example, in *Weather Prediction by Numerical Process* (1922, p. 53), he writes that "to estimate the errors due to finite differences, the calculation was repeated in exactly the same manner, but taking 6 layers of equal mass instead of 3, and dividing the hemisphere into 6 parts. . . ." And on pages 151 through 153 Richardson considers six methods of commencing the computation. To decide which to use he tries each method on a simple differential equation that has an exact solution. In the table, reproduced here as Figure 8, the values produced by each method are placed alongside the exact values. He concludes, "If we consider not merely accuracy but also ease of performance, the most satisfactory process in this case must be judged to be the one which begins with a very small uncentered step and doubles the length of the step-over several times, in the manner described in (ii) above."

For some of his equations Richardson places a small number below each term. These numbers indicate the order of magnitude of each term, so that one can decide what terms need to be kept as the equation is manipulated and fitted into the overall algorithm. It is interesting that Richardson (1922) sometimes suggests a change in an algorithm but does not test the change, presumably because of the computational labor involved.[22]

This is one of the most striking aspects of Richardson's meteorological work: the amount of time numerical analysis has center stage. There are larger roles, of course, for observational data (that, at least, of the appropriate sort) and for theories (again, those of the appropriate sort). But the large- and medium-scale phenomena of the atmosphere—prevailing winds, cyclones, cloud formations, and storms—are walk-ons, if they are cast at all. This new major player, however, all but disappeared from the stage of meteorology for some 30 years, because the one kind of production he could play in did not find backers again until then.

t time	Exact $\theta=e^{-t}$	θ by advancing time-steps of 0·2 throughout	θ by centered time-steps of 0·4, stepping over intermediate values, except at the beginning, where the following methods are employed				θ by Maclaurin as far as available
			First step of 0·2 uncentered	Small initial steps; only first uncentered	Double initial data	Algebraic first step	
		Excess of approximate over exact					
0	1·0000	·0000	·0000	·0000	·0000	·0000	·0000
0·025	·9753			−·0003			
0·05	·9512			·0000			
0·1	·9048			·0001			
0·2	·8187	−·0187	−·0187	·0003	·0000	−·0005	−·0187
0·4	·6703	−·0303	·0097	·0021	·0022	·0024	·0097
0·6	·5488	−·0368	−·0208	·0012	·0009	·0003	−·0048
0·8	·4493	−·0397	·0195	·0031	·0033	·0038	·0024
1·0	·3679	−·0402	−·0274	·0011	·0008	·0000	−·0012

Figure 8 In this table Richardson compares six different ways of commencing the computation by finding the errors at various time-steps for each of the methods. The methods are applied to the equation $d\theta/dt = -\theta$, which has an exact solution $\theta = \exp(-t)$.

The Inclusiveness of the Scheme

Perhaps the most remarkable aspect of Richardson's scheme is the *way* in which it relates dozens of phenomena to one another: the relationships are explicit, they are quantitative, and they all function simultaneously. Richardson commented, "The scheme is complicated because the atmosphere is complicated. But it has been reduced to a set of computing forms" (1922, p. xi).[23]

This aspect of Richardson's scheme was much commented on by contemporaries. The English climatologist F. J. W. Whipple, in his review of the book, wrote: "Its merit is that it insists on the study of all the various ways in which the meteorological elements act and re-act on each other. This co-ordination of knowledge has been the stimulus to many special researches valuable in themselves but appearing now, for the first time, in their proper relations" (Platzman, 1967, p. 521). The meteorologist Sydney Chapman wrote: ". . . in discussing the changes of heat energy in the atmosphere it is necessary to consider changes of volume by convergence of winds, the amount of heat reaching any particular layer by radiation . . . by transfer of water, and by eddy conduction from neighboring air or from the ground; many of these subjects have further ramifications, and give rise to extremely interesting resumes of recent research in various branches of meteorology. . . ." (1922, p. 283). Napier Shaw admitted that "we cannot offer it to the reader as Richardson's "Ready Reckoner" for forecasts," but continued, ". . . the effort to bring the processes of weather under numerical computation is by no means wasted. In the course of mapping out the computation, as in no other way, the dynamics and physics of many of the processes of weather are made clear, and a very large amount of information about the atmosphere, difficult to acquire, is contained in the book" (Platzman, 1967, p. 542).

Richardson succeeded in fitting together a great many quantitative theories of atmospheric phenomena. The range of theories he incorporated into his scheme is vast. Some examples are the theory of gases, pressure and water content as a function of height, Egnell's law, Laplace's theory of tides adapted to the atmosphere, theory of the absorption of radiation, theory of the scattering of radiation, theory of changes in soil, theory of the effect of leaves on the interchange of heat and moisture between air and soil, Shaw's theory of winds, theory of radiation in the stratosphere, and theory of temperatures in the stratosphere.[24]

By means of such theories he succeeded in bringing a great deal of experimental and observational data to bear on weather predictions, and the variety of measurements that therefore enter into Richardson's computation is also striking. These include measurements of the following: the saturation pressure of water vapor in equilibrium with water or ice as a function of temperature, the horizontal component of the acceleration of gravity, entropy-per-mass as a function of temperature, specific heats, the ratio of air to water in a cloud, eddy viscosity, shearing stress on the earth's surface, vertical variation of pressure, radiation emission in the stratosphere, and temperature gradients in the stratosphere.[25]

It is likely that Richardson's scheme was unprecedented in all of science in the variety of phenomena represented in a single computation.[26] If any one of the component theories is changed, then one or more of Richardson's equations changes its form. If different values are found for any of the quantities measured, then one or more constants in Richardson's equations changes its value. Of course, if the initial values are changed, the predictions are changed. The theories and the data comprising the scheme are subject to separate testing[27] and to being replaced when something better becomes available.

Richardson's scheme, then, includes a great many phenomena. It also excludes a great many. The scheme both coheres and repels: either a phenomenon is tied quantitatively to everything else in the scheme, or it is left out altogether. Richardson decides, for example, to exclude the atmosphere more than some 50 km up. He says, "By this convention we free ourselves from the necessity for entering into difficult questions concerning that outer atmosphere which is ionized, which may be escaping, and in which the variation of gravity, and the term $2m_E/a$ in the equation of continuity of mass, would cause mathematical difficulties. By this convention also we assume that whatever the rare gas above h_0 may do, it has no influence on the surface weather" (Richardson, 1922, p. 125).

For providing guidance for observers and theorists, Richardson's scheme had the important property that the carrying out of a computation allows one to judge the relative importance of different effects. For example, one need only look at the computing form that sums the various terms contributing to change in the water-content to see what terms are most important. Shaw remarked that the scheme "will not only provide an acid test of meteorological theory but also be a valuable guide to the organisation of new meteorological observations" (Shaw, 1922, p. 765).

In most of the book *Weather Prediction by Numerical Process* it is the weather that Richardson (1922) treats computationally. In Section 3 of Chapter 3 he takes a step back, as it were, and treats computationally his entire scheme of weather prediction. The atmosphere, we recall, is represented by the values of seven variables (three components of velocity, temperature, pressure, density, and moisture) at the points of a four-dimensional grid (three spatial dimensions and the time dimension). Suppose we increase by a factor of n the density of grid points in each dimension. (For $n = 2$, each square is half as high and half as wide, and observations are made at twice as many heights and twice as often.)

Richardson reasons as follows. Since the number of grid points increases by a factor of n^4, the monetary cost of doing the computation increases by that factor. The number of squares, hence the number of observing stations, increases by a factor of n^2. The cost at each station would be little affected by increasing the number of observations to be made there (though some additional observers would probably be required), so it is reasonable to assume that the cost at each station increases by a factor of n (rather than a factor of n^2, by which the number of observations made there increases). The administrative cost for the entire scheme is assumed to be the constant amount c. Then the total cost C is given by

$$C = a\,n^4 + b\,n^3 + c,$$

where the first term on the right is the cost of computing and the second term is the cost of gathering the data (a and b are constants of proportionality). Notice that as n increases the cost of computation comes to dominate total cost regardless of the values of the constants.[28]

Richardson does not give a quantitative expression for the benefits of his scheme,[29] but he does relate the cost of the scheme to improvements in accuracy. He refers to his 1910 paper in which he showed that for small finite-differences, the errors are proportional to $1/n^2$. "Thus as n varied, *the errors would be inversely as the square root of the cost of computing alone*" (Richardson, 1922, p. 18). Hence, halving the error requires quadrupling the amount paid for computation.

It is interesting to see how readily a computational model of the atmosphere itself serves as the object of computational analysis. This is not true, in general, of qualitative models of the atmosphere, nor of mathematical models that are not computational. Such analysis had, however, little or nothing to do with the fate of Richardson's scheme. What the main determinants were is discussed below.

The Influence of Richardson's Work

Richardson's work encouraged a few meteorologists to take a computational approach to particular phenomena, because he showed how partial differential equations that arise in meteorology can be solved arithmetically. An important paper published by William H. Dines in 1920, "Atmospheric and terrestrial radiation," is based on Richardson's method of calculation (Platzman, 1967, p. 540). As we will see in Chapter 8, computing forms similar to Richardson's were used by a number of meteorologists. Vilhelm Bjerknes and the rest of the Bergen meteorologists continued their efforts to derive meteorological results from the laws of physics, but they held to graphical methods to solve equations.

The Cambridge University Press published separately the set of computing forms that embody Richardson's scheme. It appears that no one—not even Richardson[30]—ever used these to make another test of the scheme. Moreover, it appears that no one—other than, to a small extent, Richardson—ever tried to modify the scheme to make it either more practical or more accurate. Richardson wrote wistfully, "Perhaps some day in the dim future it will be possible to advance the computations faster than the weather advances and at a cost less than the saving to mankind due to the information gained. But that is a dream" (1922, p. xi).

But *Weather Prediction by Numerical Process* was not ignored. It was immediately reviewed in major journals, including *Quarterly Journal of the Royal Meteorological Society, Meteorologische Zeitschrift, Philosophical Magazine, Geographical Review, Nature, Meteorological Magazine,* and *Monthly Weather Review,* by leading scientists, including Felix Exner, Napier Shaw, and Harold Jeffreys. Shaw even wrote a new chapter devoted to Richardson's method for the second edition, published in 1923, of *Forecasting Weather.* Most of the reviewers praised the book, although only a few were enthusiastic about Richardson's approach. A number of laudatory remarks concerning the range of phenomena Richardson dealt with were quoted earlier.

It is possible that the carefulness and completeness with which Richardson constructed his scheme and the ingenuity of his computational procedures suggested to most meteorologists that they could hardly have done better themselves. This, combined with the bad prediction, must have convinced all meteorologists that a computational approach to weather prediction was completely impractical. Shaw wrote, ". . . forecasting by numerical process seems so arduous and so disappointing in the first attempts that the result is a sense of warning rather than attraction" (1922, p. 764). In 1959 Tor Bergeron wrote, "Unfortunately, Richardson's unsuccessful trial has till now withheld theoreticians within Meteorology from making a renewed attack on our main problem along the lines of this forecasting method . . . ," and Rossby commented, "After Richardson's experiment in 1922 little thought was given to the idea of integrating the general atmospheric equations of motion through numerical hand computations" (Bergeron, 1959, p. 454; Rossby, 1957, p. 31).

In his inaugural lecture in 1913 at the new institute of geophysics in Leipzig, Bjerknes said the following: "What is it that I really seek? Whither am I steering? I could not free myself from the thought that 'There is after all but one problem worth attacking, viz, the precalculation of future conditions' " (1914, p. 14). Many years later Jacob Bjerknes said: "I think I can say for certain that my father did consider Richardson's work as the real first step toward the fulfillment of the '1904 program.'" But he also saw it as a demonstration of the next to insurmountable difficulties looming ahead for numerical forecasting" (in Platzman, 1967, p. 549). Thus, Richardson's work convinced even Bjerknes that the final goal was so distant that the proper course was a "series of preparatory individual problems" (Bjerknes's phrase) rather than an attempt to compute the weather. Shaw's way of putting it was that *Weather Prediction by Numerical Process* "opens the way to useful exercises less stupendous than calculating the weather. . . ." (1922, p. 765).

In 1966 the meteorologist N. A. Phillips wrote: "Although the book made considerable impression when first published, it then appears to have been almost completely ignored until the late 1940's, when J. Charney and J. von Neumann began the modern era of numerical weather prediction at Princeton" (Phillips, 1966, p. 633). Such statements, although accurate, have led to a misperception of Richardson. Most people today who know of his work see it as an important piece of science out of its place in time and see Richardson as an ignored genius. Neither view is correct.[31] Far from being out of its place in time, Richardson's work was a full trial of the leading research program of his time. Far from being ignored, Richardson's work was widely noticed and highly regarded, and as a result it had a highly important effect—it directed meteorologists elsewhere. In short, Bjerknes pointed out a new road, Richardson traveled a little way down it, and his example dissuaded anyone else from going in that direction until they had electronic computers to accompany them.

Chapter 7 | The Growth of Meteorology

Meteorology in World War I

In the final weeks of 1916 the Great War seemed to be at stalemate. The advances made some months earlier by General Aleksei Bruselov against the Austrians had been so costly that the Russian armies had withdrawn to their former positions. On the Western Front, the French launched the last of the major offensives in the year-long battle of Verdun; true to pattern, the gains were small, the losses immense. Not far from Verdun, Lewis Fry Richardson was using his time off from ambulance driving to work out a forecasting algorithm. At the Geophysical Institution in Leipzig, Bjerknes, frustrated by the loss of many of his assistants and doctoral students to military service, was contemplating a return to Norway. In London, Napier Shaw was directing the work of the Meteorological Office. And in Vienna, the Director of the Central Institute for Meteorology and Geodynamics, Felix Exner, found enough time off from war work to write the preface to *Dynamische Meteorologie,* the body of which had been completed two years earlier.

Meteorologists were busier than ever. Regard for the weather had, of course, long been considered important in the conduct of war,[1] but it was not until this century that a meteorological staff became a standard element in military organization.[2] According to the English meteorologist Ernest Gold,

> The army [at the start of World War I] had little use for meteorology: the attitude of the General Staff to a deputation of representatives of science urging its importance was briefly that 'the British Army fights its battles with guns and bayonets and not with meteorology.' Mud, gas and aviation rapidly effected a change. (1945, p. 220)

Gold here points to some of the factors that made meteorological information more valuable in World War I than it had been in earlier wars: mechanized transport made armies more mobile than ever, but only when roads and weather permitted; the use of poison gas was tied to weather conditions; and the military use of aircraft called for a new range of meteorological services.[3] There were other factors. Robert A. Millikan, who during the war commanded the Meteorological and Aerological Service of the Signal Corps, wrote

> When it is remembered that the biggest element in the effectiveness of a modern army is its artillery and that the effectiveness of the artillery is dependent entirely upon these wind corrections it will be seen how incalculably valuable the work of

> the trained physicists and mathematicians [working for the Signal Corps] proved to be to the practical problems of the great war. (1920, p. 312)

In fact, temperature, humidity, and winds at various heights could all be taken into account through the use of ballistic tables and other calculating aids. Such information was used also in determining the location of enemy guns by sound ranging. Even climatological information was valuable to the combatants, on account of the geographic and temporal extent of the conflict. Most significant, though, were the improvements, since the last major European conflict (the Franco-Prussian War in 1870–1871), in forecasting ability and in communications (the speed and range of which are important both in getting information to the forecaster and in getting the forecasts to potential users).

All these factors combined to establish military meteorology. According to Robert Millikan, "Prior to 1914 a meteorological section was not considered a necessary part of the military service" (1919, p. 133). By the end of World War I all the major participants had meteorological services within the military,[4] and the amount and frequency of reported meteorological information was unprecedented. The artillery and air service of the U.S. Army, for example, was provided, every 2 h, with measurements of temperature, density, and wind speed and direction, at the surface and at various altitudes up to 5000 m (Millikan, 1919, p. 133).

There is no doubt that the Great War stimulated the growth of meteorology, as a quarter century later the next global conflict was to do in even greater measure. But between these wars the science experienced a remarkable growth. At the time of World War I, meteorology had standing as an academic discipline only in Scandinavia, Germany, and Austria, and, although it was recognized as a vocation in many countries, the number of meteorologists was small, formal training was scant, and there were few professional organizations. At the time of World War II, meteorology was a recognized academic discipline and a full-fledged profession throughout the Western world. In the interim, observational meteorology—aided by radio, teletype, punched-card machines, and greater governmental support—made great advances. So too did dynamical meteorology, the physics-based explanation of atmospheric motions; indeed, by the end of the period, especially through the work of Carl-Gustaf Rossby, it had become quite useful to forecasters. Weather forecasting changed even more as a new style of synoptic meteorology was developed in Bergen, halfway up the coast of Norway, by a group of young meteorologists under the direction of Vilhelm Bjerknes.

The Bergen School

As we have seen, weather forecasting in the 19th century and early 20th century was seldom based on explicit rules. A forecaster put the latest data onto maps and used his experience with maps of past weather to predict the next day's weather. Meteorologists were, however, not content with forecasting being "an art rather than a science," and they continually sought to specify a procedure for making a forecast. The strategy advocated by Vilhelm Bjerknes—use the known laws of hydrodynamics and thermodynamics to *calculate* the weather—struck most meteorologists as utterly impractical, and the work of Lewis Fry Richardson deepened the impression. A second strategy was to devise useful "higher-level" rules, rules specific to meteorology and not connected to physics.[5] By 1918 a good many had been advanced, but few of them were reliable or of much use in forecasting. It is ironic that it was the meteorologists at Bergen—headed by Bjerknes, the champion of the first strategy—who had the greatest success with the second strategy.

We saw in Chapter 5 that after Bjerknes's return to Norway in 1917 the exigencies of wartime led to his establishment in the following year of a forecasting service. Partly because the war had put a stop to weather telegrams from England, Bjerknes worked hard to expand the network of weather stations within Norway; by the summer of 1918 he had set up 60 new stations, and they were equipped to obtain accurate wind measurements (Friedman, 1989, p. 121). The meteorologists at Bergen—Bjerknes, his son Jacob, Tor Bergeron, Halvor Solberg, and a few others—thus had an extraordinarily rich observational basis for forecasts. In the course of the next several years, these meteorologists introduced a series of new concepts.

Among the new concepts were air mass, cold front, warm front, and occluded front.[6] The Bergen meteorologists saw the atmosphere of the northern hemisphere as polar air separated from tropical air by a so-called polar front. In their view, the cyclones of the northern temperate zone, the low-pressure areas of counterclockwise-circulating air, developed from waves in this boundary surface between tropical and polar air, and the Bergen meteorologists went on to propose a complete model of cyclone development and dissipation.

These concepts, which arose in the practice of providing forecasts, were the basis of forecasting techniques developed in Bergen. The techniques, which came to be known as air-mass analysis, received a great deal of attention and were soon adopted by individual forecasters in countries worldwide. The American meteorologist Jerome Namias wrote, "The concepts made order out of the apparent chaos of weather. They provided a practical method that the forecaster could use in his daily work" (Basu, 1984, p. 193).

But the stodginess of many national weather services delayed by a decade or so their general adoption. For example, fronts were not drawn on the weather maps published in the English *Daily Weather Report* until 1933 (Douglas, 1952, p. 9).[7] The U.S. Weather Bureau did not begin making use of air-mass analysis until 1934, and it was not until 1936 that fronts were drawn on many of the maps prepared by the Weather Bureau (Whitnah, 1961, p. 161).[8] These were, however, among the last holdouts. In 1933 David Brunt reported that, "The Norwegian school of thought has attained almost complete acceptance by the whole world of meteorology . . ." (p. 96), and by the end of the decade it was generally accepted that the Bergen methods improved forecasting substantially, particularly the forecasting of precipitation.

The Growth of Dynamical Meteorology

The success of air-mass analysis did not, however, stop the development of physics-based meteorology. Bjerknes too contributed, as he worked to connect the empirical theory and physics. The persistence of the view that meteorological theory must be based on physics is seen in a statement made by the English meteorologist C. K. M. Douglas:

> The Norwegian work was often referred to as the 'Bjerknes theory' but the word 'theory' is unsuitable for anything in synoptic meteorology. It is really a technique based on simplified models of atmospheric structure and movement, and its successes have been based on its empirical rather than its theoretical aspects. (1952, p. 6)

Douglas (1952, p. 9) argued that the Bergen techniques were even more subjective than earlier techniques, in the sense that there was greater variability, from forecaster to forecaster, in the analysis of weather maps. So alongside the largely nonmathematical air-mass analysis of the Bergen School, dynamical meteorology continued its growth. Both this growth and resistance to it are suggested in the following sentences from the preface to the second edition, published in 1939, of David Brunt's *Physical and Dynamical Meteorology:*

> A few reviewers of the first edition [published in 1934] complained of the number of equations it contained. I make no apology for having perhaps added to the number of equations, as I take the view that meteorology should aim at being a metric science wherever possible, and that no physical theory can be regarded as wholly satisfactory which cannot be expressed in mathematical form.[9]

The growth of dynamical meteorology owed much to the gradual establishment of meteorology as an academic discipline. At the turn of the century there were professorships in meteorology in Germany, Austria, and Scandinavia, but it was not until 1920 that the first professorship in meteorology in the British Commonwealth was established, when Shaw was given a chair at Imperial College, London.[10] In 1901 the *Monthly Weather Review* reported on the situation in the United States[11]:

> In general, meteorology continues to labor under the disadvantage of failing to secure distinct and independent recognition in our colleges and universities. Some treat it as a small branch of geography, others as belonging to geology; many class it with the mathematical and experimental physics; in a few cases it keeps its ancient association with chemistry and natural philosophy. (Vol. 29, p. 264)

And in 1906 Cleveland Abbe complained, "Meteorology is not yet properly recognized in our colleges, nor as a postgraduate course in our universities" (1907, p. 309). Finally in 1928 the Massachusetts Institute of Technology established the first professional-training program in meteorology in the United States.[12] In 1940 meteorology became a separate department at MIT, and by the end of that year there were departments of meteorology in four other universities in the United States.

Meteorology as a Profession

It was also in the interwar period that meteorology firmly established itself as a profession. The main impetus came from the interest in aviation—military aviation during the war and commercial aviation afterward. The meteorological requirements of aviation were a major issue at the Paris Peace Conference, and international standards for meteorological observations and forecasts were there adopted. Most of the increased expenditure for the U.S. Weather Bureau in the 1920s went to the provision of meteorological service for aviation (Whitnah, 1961, p. 181).[13]

Alhough aviation was most important, other factors contributed to the expansion of meteorology. The experience of World War I had made the military services more interested in meteorology. Automobile travel and a marked increase in ocean travel made people more interested in weather forecasts. There was a new belief that hurricanes and tornadoes could be predicted. Abundant upper-air data became available for the first time because of airplane observations, beginning during World War I, and radiosondes,[14] beginning in the late 1920s. Radio, from 1920 on, and the teletype, from 1928 on, facilitated the transmission of weather information enormously. And, partly as a result of the successful forecasting of the Bergen meteorologists, there was a new optimism about forecasting in general.

As a result of all these things many governments sharply increased the funding of weather services. In 1926 the British government spent eight times as much on meteorology annually as it did before the war, and annual appropriations for the U.S. Weather Bureau climbed from less than $2 million just after the war to $4.5 million in 1932 (Shaw, 1926, p. 2; Whitnah, 1961, p. 21). The airline companies hired quite a few meteorologists (Basu, 1984, p. 94), and there began to be a wider demand for meteorological consultants. One of the most successful was the American meteorologist John P. Finley, who in the 1920s gave advice to insurance companies concerning the risk of damage from tornadoes, windstorms, and hail for different areas of the country (Galway, 1985, p. 1509).

In the late 19th century two organizations for meteorologists in the United States, one started by Cleveland Abbe and the other by Robert De Courcy Ward, proved short-lived (Brooks, 1950). A sign of the emergence of a profession of meteorology in this period is the vitality of the American Meteorological Society. It was founded in 1919, and by 1940 its membership had doubled.[15] Also significant is the fact that the International Meteorological Organization, after operating informally for 50 years, established a permanent office (in Holland) in 1931. In 1929 this organization began a worldwide standardization of meteorological codes, units, and symbols. Six years later Willis Gregg wrote, "It can be said, without fear of contradiction, that the accomplishments of this organization in bringing about uniformity of practice have no parallel, either in scientific endeavor or in the fields of international politics and commerce" (1935, p. 339).

Something of the status of meteorology is apparent from a listing of the most-printed meteorological books. From the middle of the 19th century onward there were a great many published compilations of data and a considerable number of manuals and handbooks for observers. Yet until 1917 there were very few comprehensive presentations of meteorological theory. Between 1917 and 1939 dozens of such books were published. Among the most important were Exner's *Dynamische Meteorologie* (1917), Humphreys's *Physics of the Air* (1920), Richardson's *Weather Prediction by Numerical Process* (1922), the four volumes of Shaw's *Manual of Meteorology* (1926, 1928, 1930, 1931), Vilhelm Bjerknes *et al.*'s *Physikalische Hydrodynamik* (1933), and Brunt's *Physical and Dynamical Meteorology* (1934). Other important books were those by McAdie (1917), Lempfert (1920), Baldit (1921), Geddes (1921), Clayton (1923), and Byers (1937). A sign of the establishment of meteorology as a profession is the publication of books having such titles as *Meteorology as a Career.*[16]

Carl-Gustaf Rossby

In 1933 Antonio Giao wrote, "It is indeed certain that for meteorologists there is a radical separation between theory and forecasting" (1935, p. 77).[17] In 1946 H. G. Houghton wrote that "Our physical understanding of atmospheric processes is so limited that it is of little utility in weather forecasting" (in Douglas, 1952, p. 16). These are overstatements: Shaw's tephigram, Bell's gradient-velocity nomogram (shown in Chapter 8), and Petterssen's (1939) fog-prediction diagram are examples of forecasting tools based on theory. But there is no doubt that in the interwar period most forecasters did not see dynamical meteorology as having much relevance to their work. The man who did most to change this perception, Carl-Gustaf Rossby, explained its accuracy in 1934:

> One of the greatest obstacles to progress in meteorology is undoubtedly to be found in the wide gulf between the mathematical theory on the one hand and the applied science on the other. . . . meteorological theory in many cases has degenerated into pretty pieces of mathematical exhibition, where the postulates lack all resemblance to the conditions actually found in the air and where the results can not be checked. . . . This . . . has caused such a deep distrust, particularly in the United States, of theoretical investigations that synoptic meteorologists have restricted themselves to an accumulation of weather-map experience which is seldom or never interpreted except in the most superficial sense. (pp. 265–266)

Born in Stockholm in 1898, Carl-Gustaf Rossby specialized in mathematical physics at the University of Stockholm (Stockholms Högskola), from which he graduated in 1918. He worked for 2 years at Bjerknes' institute in Bergen and then returned to the University of Stockholm to learn more mathematical physics, receiving a licentiate in 1925. He was then granted a fellowship to study at the U.S. Weather Bureau in Washington. Rossby stayed in the United States for some 25 years.

Horace Byers has written: "Rossby was really two men. On the one hand he was the organizer, director, and promoter and on the other the scholarly research scientist" (1960, p. 249; see also Byers, 1959). Rossby's career as researcher had already begun, with an article written in 1923. His career as organizer began in 1928 when he was chosen to set up a weather service for a trial airway service between Los Angeles and San Francisco. Rossby's became the model for the weather services of commercial airlines.

To convey the magnitude of Rossby's work as organizer a bare listing must here suffice. He was largely responsible for the establishment of meteorology programs at MIT (begun in 1928) and at the University of Chicago (begun in 1940) and for the reorganization of the meteorological institute at the University of Stockholm (in about 1950). He played a major role in the reorganization of the U.S. Weather Bureau in the late 1930s, making research a more important function of the Weather Bureau, and in the postwar reorganization of the American Meteorological Society. During World War II he was the leading figure in the meteorological training program of the Army Air Forces, which produced some 7000 meteorological officers (Byers, 1970, p. 215). He was partly responsible for the founding of two research journals (*Journal of Meteorology* and *Tellus*), for the establishment of a meteorology program at UCLA (begun in 1940), and for the initiation of regular computer forecasting in Sweden (in about 1955).

In several ways Rossby was like Richardson. He shared with Richardson a determination to get numerical answers out of theory: in the late 1920s Rossby (1929) devised both numerical and graphical techniques for calculating the work done on a parcel of air being displaced upward by buoyancy forces. Like Richardson, Rossby accepted one of the 20th-century's great challenges: to devise a mathematical theory of turbulence. In a 1932 paper he introduced into meteorology important concepts from aerodynamics (mixing length, roughness parameter, and von Karman's constant). Byers said that in a 1935 paper of Rossby's "the physical (as contrasted with the later statistical) approach to turbulence was carried toward perfection, although it remained a subject which could be treated only imperfectly" (Byers, 1960, p. 256).

In the late 1930s Rossby set to work to calculate, on the basis of physics, the large-scale motion of the atmosphere. Like Richardson he succeeded in getting a machine whose wheels would turn. But he was more ruthless than Richardson in simplifying—choosing to ignore friction, radiation, and the water-vapor cycle of the atmosphere—and got a more useful machine.

Rossby's two most famous papers appeared in 1939 and 1940. In the first paper he discussed certain long-wavelength waves in westerly currents (now called Rossby waves). The propagation speed of these waves he gave in an equation which, according to the historian Gisela Kutzbach, is "perhaps the most celebrated analytic solution of a dynamic equation in meteorological literature" (1975, p. 558). In the second paper he advanced the concept of constant-vorticity trajectories of winds and showed how to use it to calculate air movement. On the basis of these results Rossby and his collaborators in 1940 made numerical predictions for a one-layer atmosphere.

Rossby's work was significant because the equations both fit the observations well and could be solved. The calculation of the propagation speed of Rossby waves could be done with an ordinary slide rule. The calculations of constant-vorticity trajectories, however, were difficult enough that meteorologists soon devised calculating aids for this purpose: in 1943 a slide-rule system (part of which is pictured in Chapter 8) by J. C. Bellamy; in 1945 a set of tables by S. Hess and S. Fomenko; and in 1951 a mechanical differential analyzer by H. Wobus (Godske *et al.,* 1957, pp. 715–719).

In 1922 Richardson showed how a forecast might be based on meteorological theory. According to G. P. Cressman: "The next significant move in the field of dynamic prediction came in 1939, seventeen years after the publication of Richardson's book, when Rossby (1939) published his well known exposition on the movement of long waves in a westerly current" (Cressman, 1972, p. 182). This too is an overstatement, but Rossby was no doubt the key figure in the rapprochement of forecasting practice and physics-based theory. Rossby's work, according to Jule Charney, "injected new vitality into dynamic meteorology. For the first time, a dynamic theory was presented in which the characteristic 'planetary' properties of the atmosphere were taken into account" (Charney, 1950, p. 234). Rossby's great achievement was to devise a theory that was based on physics, that described atmospheric phenomena that were of importance to forecasters, and whose predictions could be *calculated.* We will see in Chapter 9 how the wartime increase in the gathering of upper-air data made Rossby's theory even more useful, and in Chapter 10 how it served as the starting point for von Neumann's Meteorology Project, which finally fulfilled Bjerknes's program of calculating the weather.

Chapter 8 Meteorological Calculation in the Interwar Period

During the 1920s and 1930s calculation assumed a much larger role in meteorology. The amount of data processing increased sharply, although less because of more sophisticated processing than because of the intensification of data gathering. More people than ever before were studying dynamical meteorology, and they continually sought to connect theory and data, hence were continually doing calculations. Weather forecasting remained, on the whole, nonquantitative and noncalculational, but a number of types of predictions could, for the first time, be arrived at by calculation. Because of the calculations involved in data processing, in dynamical meteorology, and in forecasting, there was during the interwar period a remarkable proliferation of calculating aids. We look first at a class of devices whose adoption by meteorologists was clearly the result of data push.

The First Use of Punched-Card Machines

In respect of quantitative observational data, meteorologists always were overwhelmed with what they had and, at the same time, worked incessantly to acquire more. In the 1920s and 1930s the disproportion between the accumulation of data and what had been done with the data became somewhat of an embarrassment. In 1932 the President of the Royal Meteorological Society R. G. K. Lempfert wrote

> When a meteorologist surveys his library he can hardly fail to experience a sense of uneasiness at its ever-increasing bulk. . . . Text books and treatises form only a small proportion of the books on the shelves. The great majority of them contain nothing but meteorological data. (p. 91)

More countries were making reliable observations, and the international exchange of data increased. The countries that had long collected data increased the number of weather stations they maintained. There were new ways of gathering data, notably by airplane and radiosonde. New ways of communicating led to more data being reported; most important was the use of radios on ships, which gave forecasters an abundance of observations from a region never before contributing timely data. There were new reasons for collecting data: for aviation, for forest-fire control, for automobile travel, for drought amelioration, and for navigation in northern waters (Hughes, 1970, pp. 44, 48, 69, 107).

In the late 1880s the U.S. Census Bureau faced a crisis of a similar sort. The tabulating and analyzing of the data from the 1880 census was dragging on, to be finally completed in 1889. It was obvious that the U.S. population was growing rapidly, and there was demand both for more data from each household and for more analysis of the data. The need for new methods being urgent, the Census Bureau conducted a trial of three new systems. The clear winner was the system of card-punching, card-sorting, and card-tabulating machines designed by Herman Hollerith, and the Census Bureau immediately acquired 56 sorters and tabulators.

Hollerith's machines performed well with the 1890 census. The total population (62,979,766) was announced just 6 weeks after the count began. Although the complete analysis took almost 7 years and cost almost twice as much as the analysis of the 1880 census, much more was done with the data than ever before. In 1891 Robert P. Porter of the Census Bureau said, "Because the electrical tabulating system of Mr. Hollerith permitted easy counting, certain questions were asked for the first time" (Goldstine, 1972, p. 69). Hollerith's machines were soon in use for census purposes all over the Western world, but, with one exception, they were not used for meteorological data until the 1920s.

The exception was the U.S. Navy Hydrographic Office. In 1838 it had been directed by the Secretary of the Navy to collect a continuous series of meteorological observations taken every 3 h day and night. In 1893, when a new director, C. D. Sigsbee, was appointed, there were more than 3000 observers sending data to the Hydrographic Office, and the flow of data could no longer be properly processed. Sigsbee wrote in the annual report in 1895:

> Investigation was made of the system for electrical counting and averaging invented by Mr. Herman Hollerith for use in the last census of the United States. It was ascertained that this rapid, accurate and economical system could be utilized for the work of the office. Much study has been given to the question in order to cover possibilities of personal error in filing, handling, and recording, and in order that steps once taken need not be retraced. Considerable progress has been made, and it is hoped that the system will soon be in complete operation. (in Bates, 1956, p. 521)

This use of punched cards and tabulating equipment was apparently short-lived: in 1904 President Theodore Roosevelt signed an order abolishing the Division of Marine Meteorology of the Hydrographic Office and transferring its records to the Weather Bureau (Bates, 1956, pp. 519–522).

In about 1920 the Meteorological Office of the British Admiralty began using punched-card methods to compute summary statistics. In 1922 the Dutch Meteorological Institute, having borrowed some British card-files, began using punched-card machines. So did Norway, France, and Germany soon thereafter. In the mid 1920s the Czechoslovakian meteorologist L. W. Pollak designed an inexpensive punch machine and had one placed in every Czechoslovakian weather station, and by 1927 Pollak had used the tabulating machines to produce frequency tables of barometric pressure. In the late 1930s the weather service of every major European country was analyzing data by means of punched cards (U.S. Department of the Air Force, Air Weather Service, 1949, p. 1; Conrad and Pollak, 1950, p. 351).

A laggard in this movement was the weather service of the United States, where the Hollerith method had originated. The U.S. Weather Bureau had considered acquiring some of Hollerith's punching and tabulating machines in about 1885 and in 1895 (Austrian, 1982, p. 112; Whitnah, 1961, p. 66), but it was not until the mid 1930s that the Weather Bureau actually did so. In 1934 a Science Advisory Board recommended a card-punching unit for the central office of the Weather Bureau, and funds were provided for card-punching by two of President Franklin Roosevelt's programs to ameliorate the Depression—by the Civil Works Administration in 1934 and by the Works Progress Administration in 1936 (Whitnah, 1961, p. 157).[1] A number of meteorological atlases were produced using these cards, such as *Atlas of Climatic Charts of the Oceans* (1938) and *Airway Meteorological Atlas for the United States* (1941). For the production of the latter, more than 14 million airway observations were transferred to punched cards (Conrad and Pollak, 1950, p. 353).

Tabulating equipment thus made it possible to use many more data than was possible before. Another important result was a higher standard of weather data. Such machines as duplicating punches, reproducing punches, verifiers, and interpreters[2] eliminated most errors from a great many of the routine data-processing tasks. More errors were eliminated when sorting and tabulating were done by machine. And machines were used to check for missing data and to identify obviously erroneous data.

A third effect of the use of punched-card machines was the facilitation of calculation, especially for sophisticated statistical analyses. Originally these machines were used only for sorting, searching, and counting. As the tabulating equipment became more sophisticated—IBM introduced in 1931 a multiplying punch and in about 1933 removable control panels—meteorologists were able to do more and more complicated computations mechanically (Bashe *et al.*, 1986, p. 17).

It was in this period that for the first time a great many people applied statistical techniques, beyond simple averaging, to meteorological data. Statistics was used in weather forecasting (discussed below). It was used in hypothesis testing, as in a 1922 article by Richardson, Wagner, and Dietzius, or as in the search for weather cycles (also discussed below.) It was used to reveal patterns in the data, as in F. J. W. Whipple's 1924 article on regression equations in the analysis of upper-air observations (1924b), or in T. N. Hoblyn's 1928 study of extreme temperatures.

One of the reasons statistics came to be used more commonly in the interwar period was that, for the first time, mechanical calculators became widely available to meteorologists. It was in the early decades of this century that mechanical calculators became widely used, both in the commercial and the scientific world. Among the most popular machines of the 1920s were Felt's Comptometer, Odhner's Calculating Machine (patented in 1891), the Brunsviga (some 20,000 were sold between 1892 and 1912), Steiger's Millionaire (some 5000 were sold between 1899 and 1935), the Mercedes–Euklid Calculating Machine (first marketed in 1910), and various Monroe calculating machines. Many of these machines were used by meteorologists. The American meteorologist Daniel Draper used various calculating machines, including a Comptometer in about 1890, and Richardson used both the Odhner and the Mercedes–Euklid (Ashford, 1985, p. 246; Richardson, 1922, p. 13).

The computational labor involved in the use of statistical methods was a major hindrance to their acceptance by meteorologists. A paper read in 1925 before the Royal Meteorological Society, which included the observation that the calculation of a correlation coefficient could take an hour, gave two reasons for the fact that meteorologists had until then made relatively little use of statistics: one was asked to do a great deal of arithmetic, which was exceedingly time-consuming even with the aid of calculators, and the results were "often not convincing, for some workers are continually finding small coefficients indicating relationships which break down as soon as they are applied to forecasting" (Walker and Bliss, 1926, p. 73). The tedium and the unimpressive results hardly endeared statistical analysis to meteorologists, and it remained a minor methodology and one that was looked on with suspicion by many meteorologists. In 1932 Lempfert wrote, "As the figures pass through the mill month by month, one cannot suppress the thought that the time spent in compiling such statistics could be more usefully spent in other ways." (p. 97)

The Search for Weather Cycles

One area in which statistical analysis was widely adopted—and with a decisive, although negative, effect—was in the search for weather cycles. This search—like the search for the philosopher's stone or for a proof to Fermat's Last Theorem—deserves a prominent place in the history of science. The search for weather cycles, which began in antiquity and continues today, has very often been carried out, not by meteorologists, but by people fascinated by cycles in general.[3] Most meteorologists abandoned the search in the 1930s, and calculating aids played an important part in this defection.[4]

There are obvious annual and diurnal cycles in the weather, and other cycles have been reported continually over the past 2000 years. The Biblical story of Joseph's interpretation of the Pharaoh's dream—7 fruitful years followed by 7 lean years—may reflect early belief in weather cycles. Pliny, in his *Natural History,* attributes to Eudoxus the discovery of the regular recurrence of meteorological phenomena every 4 years (Frisinger, 1977, p. 10). Francis Bacon mentions a 35-year weather cycle (Shaw, 1926, p. 118).[5] In the 19th century the periodic changes of barometric pressure caused by the sun and the moon, the so-called atmospheric tides, were detected, and the sunspot cycle of approximately 11 years was discovered. A. Kh. Khrgian, in his history of meteorology, writes "It would be impossible to mention here the names of all those who, in the 19th century, wished to find an effect of sunspots on the weather or climate" (Khrgian, 1959, p. 323).[6] And, according to Richard Gregory, "Equally imposing in its bulk with that of the sunspot cycle is the literature of cycles of two to five years" (1930, p. 115).[7]

In Volume 2 of his *Manual of Meteorology* Napier Shaw listed some 200 weather cycles empirically discovered; some 120 different cycle lengths are reported, ranging from 1 year to 260 years (1928, pp. 320–24). It appears that this search for weather cycles was most intense in the 1920s.[8] Shaw wrote that ". . . the computation of correlations [in the search for periodicities] has been taken up *con amore* and even more than that, both in the old world and in the new" (1928, p. 330). To give some impression of research in weather cycles in the 1920s, Table I lists the weather cycles mentioned in a review article on weather cycles published in 1930.

Yet by the 1940s this activity was abandoned by almost all meteorologists. This probably would not have happened had the search for weather cycles not become highly computational. Earlier, computation had played only a small role. Perhaps the most common procedure was to make a graph of some meteorological element, such as temperature or rainfall, as a function of time and then to examine the graph for some pattern in the peaks and valleys. The fecundity of this method, however, led to skepticism, and people sought a more rigorous method of detecting periodicities.

This was provided in 1897 by Arthur Schuster. In a paper entitled "On the investigation of hidden periodicities with application to a supposed 26 day period of meteorological phenomena," Schuster wrote

> It is the object of this paper to introduce a little more scientific precision into the treatment of problems which involve hidden periodicities, and to apply the theory of probability in such a way that we may be able to assign a definite number for the probability that the effects found by means of the usual methods are real. . . .

Schuster and other investigators refined the procedures presented in this paper and proposed new ones. The result was that the search for weather cycles became more and more computational. In 1926 Shaw said

> We revert to the methods which statisticians have developed to guard against a false impression of that kind [when the visual comparison of graphs suggests relationships that cannot be rigorously defended]. They deal with the numbers themselves, thus denying any opportunity for the deception, willing or unwilling, of the eye. (p. 280)

The computations, such as the computation of correlation coefficients or of Fourier coefficients, were tedious in the extreme. The tedium is testified to directly and indirectly, as in thanking a university research committee for a grant

Table I

The Weather Cycles Mentioned in a 1930 Review Article by Richard Gregory Entitled *Weather Recurrences and Weather Cycles*

Date	Investigator	Phenomena	Length of Cycle
1912	A. Defant	Rainfall	approx. 7, 12, 17, 31 days in Argentina and Australia; 5.7, 8.7, 12.7, 24.5 days in USA, Europe, and Japan
1929	J. R. Ashworth	Rainfall	below average rainfall on Sunday
1924	L. Weickmann	Pressure	pressure curve in an interval symmetric about a point
1927	Milward	Storms	tendency to occur in the middle of the month
1869	Alexander Buchan	Temperature	6 cold and 3 warm periods each year at specified dates
1896	H. C. Russell	Droughts	19 years
1890	Eduard Brueckner	Rainfall, pressure, and temperature	25 historical cycles, average length 34.8 years (ranging from 20 to 50 years)
1914	Henry Ludwell Moore	Rainfall	8 and 33 years
1918	C. Easton	Winter severity	89 years
1880	E. J. Lowe	Droughts and frosts	11 years, maximum variability at the end of each century
1923	Meterological Office, London	Rainfall	11 years
1909–10	G. Hellmann	Rainfall	2 maxima and 2 minima in each 11-year cycle
1900–01	Alexander Buchan	Rainfall	11 years
1913	A. P. Jenkins	Rainfall	3 years
1925	J. Baxendell	General weather	5.1 years
1923	Carle Salter and J. Glasspoole	Rainfall	2, 5 years
1929	Ernst Rietschel	Winter temperature	2.2 and 3-3.5 years
1924	Dinsmore Alter	Rainfall	10, 15-16 years
1927	Dinsmore Alter	Rainfall	24½, 41, 51 years
1919	C. Braak	Rainfall	3 years
1927	H. P. Berlage	Rainfall	3 years
1927	L. Petitjean	Rainfall	6, 15, 35 years, rainfall curve symmetric about 1903
1929	H. A. Hunt	Rainfall and temperature	4 years

under which part of the computing was done.[9] The number of calculating aids that were put to use is also evidence of the amount of computational labor involved. Table II is a list of some of the calculating aids used by meteorologists in the effort to detect periodicities.[10] It is noteworthy that most of them are analog devices.

In the course of the 1920s and 1930s more and more investigators reported negative results in the search for periodicities. For example, Dinsmore Alter concluded one paper as follows: "These results from widely separated parts of the world seem to show definitely that a simple 11-year period does not exist" (1924, p. 483). He added, "Nothing more is definitely shown here. It is probable that one or more periods exist in the neighborhood of 10 years," which shows that he still expected to find periodicities. The review article by Gregory, which was the source of the information in Table I, mentioned some studies that reported no appreciable periodicities: C. Cree's 1924 study of temperature, rainfall, cloudiness, and sunshine; F. Baur's 1925 study of temperature; Brunt's 1927 study of temperature; C. J. P. Cave's 1927 study of the general weather; Brooks's 1928 study of pressure. It is notable that all of these studies were done in the 1920s.

There were many users of graphical methods of finding periodicities who were reluctant to accept the judgments given by numerical tests, and this group included Shaw (1926, p. 286). The prominent American meteorologist Frank H. Bigelow complained that the statistical method "always leads to zero results in dealing with solar and terrestrial phenomena," whereas the visual method "offers some hope of success" (Walker, 1925a, p. 342).

But more and more meteorologists accepted the results of the statistical tests and became skeptical of the reality of weather cycles. In 1935 Willis Gregg reported that "effort along this line has been largely abandoned in most countries, the exceptions being Germany and the Union of Soviet Socialist Republics" (p. 341). And in 1950 V. Conrad and L. W. Pollak wrote, "Apart from the daily and annual periods of meteorological elements, the physical reality of which is evident, up to the present it has not been possible to prove by Schuster's method and his criterion that a single meteorological period is significant. . . ." (p. 411). The search for weather cycles would not have ended without numerical tests for the reality of periodicities and without abundant data. And the end would not have come so suddenly had there been no calculating aids to use in applying the tests to the data.

One of the few scientists of the 1930s and 1940s who persisted in the search for weather cycles and for a connection between solar phenomena and terrestrial weather was John Mauchly. Mauchly, born in 1907, began working with geophysical data at an early age.[11] His father, S. J. Mauchly, headed the Terrestrial Electricity and Magnetism Section of the Carnegie Institution in Washington. While a high-school student John spent many hours with a Millionaire calculator helping his father carry out computations with data on the electrical field in the earth's atmosphere. Scientists knew that solar activity affected the earth's magnetic field, and in the mid 1930s, shortly after completing a Ph.D. in physics, Mauchly set out to show that solar activity affected the weather as well. The pre-

Table II

A Listing of Some of the Calculating Aids Used in the Search for Weather Cycles, with Indication of Whether They Were Invented for This Purpose

Device	User	Description	Invented for the search for weather cycles?	Source
Buys-Ballot schedule	Buys-Ballot	Tabular and graphical procedure for finding amplitude and phase of a periodic phenomenon	Yes	Buys-Ballot (1847)
Devices to facilitate the making of Buys-Ballot schedules	G. Darwin	Device made up of celluloid strips	Yes	Pollak (1925)
	D. Gibb	Device made up of wooden cubes	Yes	Pollak (1925)
	L. W. Pollak	Automatic machine using motion-picture film and photographic paper	Yes	Pollak (1925)
	L. W. Pollak	Punched-card equipment	No	Pollak (1925)
Schuster's periodogram	A. Schuster	Graphical procedure for determining hidden periodicities	Yes	Schuster (1900)
Mader's harmonic analyzer	O. Mader	Mechanical device incorporating a polar planimeter	No(?)	Mader (1909)
Pollak's Rechentafeln	L. W. Pollak	Tables for doing harmonic analysis	Yes	Pollak (1926)
Harmonic dial	F. J. W. Whipple	Graphical method of comparing the amplitude and phase angle of periodic phenomena of the same period	Yes	Whipple (1917)
Fuhrich's self-correlating method	J. Fuhrich	Tabular procedure for finding periodicities in the order of their importance	Yes	Fuhrich (1933)
Devices to facilitate Fuhrich's method	L. W. Pollak and F. Kaiser	Punched-card equipment	No	Pollak and Kaiser (1934)
	L. W. Pollak and F. Kaiser	Tabular, graphical, and other calculating aids	Yes	Pollak and Hanel (1935)
Rotary periodograph	G. B. Moncrieff-Yates	Photoelectric device for the analysis of a disturbed periodic-curve	No	Moncrieff-Yates (1947)

vailing skepticism toward this endeavor did not dissuade Mauchly; in fact, he said, it spurred him on.[12]

Mauchly began working with the meteorological data in order to demonstrate statistically a correlation between solar activity and the weather. In particular he sought to show, since the rotational period of the sun was approximately 27 days, that there was a 27-day periodicity in the weather. Even with the help of a dozen students (paid for by the National Youth Administration, another New Deal program) and several Marchant calculators, the work went too slowly to satisfy Mauchly. So in 1939 and 1940 he built an electrical harmonic analyzer, which was an analog computer, the quantities being represented by voltages. This device sped up the calculation of harmonic coefficients by a factor of 5 or 10, and the accuracy was just sufficient—to two significant figures—for the purpose. With this device Mauchly detected periodicities in rainfall of $13\frac{1}{2}$ and 27 days.

Mauchly made plans to build a more elaborate harmonic analyzer, but then sometime in 1941 abandoned these plans to work on a digital electronic device, apparently because of the greater accuracy possible with a digital device. And from this time on Mauchly gave almost all of his time, not to the analysis of meteorological data, but to the design of electronic digital computers. His success is well known. He and J. Presper Eckert were principal designers of four famous computers: the ENIAC, the EDVAC, the BINAC, and the UNIVAC.[13] But Mauchly apparently never published, either before 1941 or after, any of the results of his study of meteorological data (Burks and Burks, 1988, p. 103).[14] Thus Mauchly contributed little to meteorology; it was rather meteorology, or meteorological data, that contributed Mauchly to the computer field.

Calculating Aids

As we have seen, the press of burgeoning data moved climatologists to begin using punched-card machines and Mauchly to devise an analog computer. Forecasters too felt data push. The initial processing of data, before they were put on weather maps, consumed much labor, especially the conversion of units of measure. International agreement, just before World War I, to use the millibar as the only unit for pressure in the international exchange of data reduced the labor considerably. Ernest Gold of the British Meteorological Office wrote, "The economic advantage [of the agreement to use the millibar] through the elimination of the conversion of thousands of values in hundreds of meteorological offices daily is substantial" (1945, p. 214).[15] But temperatures, elevations, speeds, and humidity readings were still recorded in different units in different countries. And initial processing involved more than conversion of units: barometric pressures were converted to equivalent sea-level pressure, potential temperatures were calculated, and hygrometric information was required in various forms. A variety of calculating aids expedited these tasks.

Although the use of tables may have declined somewhat relative to other ways of calculating, many new tables were devised. Some examples are the tables presented by T. N. Doerr in 1921, by G. C. Simpson in 1929, and by J. C. Ballard in 1931. Computing forms, of the sort used by Richardson in his *Weather Prediction by Numerical Process,* were used by many other meteorologists, among them Sachindra Nath Sen in 1924, G. C. Simpson in 1929, and Samuel B. Solot in 1939. Calculating machines came to be used more commonly. Certainly the use of mechanical calculators increased; we saw that Richardson and Mauchly made use of them, and many of the meteorologists doing statistical analyses used calculators. Punched-card machines were, as mentioned above, used to carry out computations. The 1930s saw the development of Vannevar Bush's differential analyzer, an electromechanical analog computer; it was used by meteorologists in the 1940s, and perhaps earlier.

Most conspicuous in the meteorological literature are the many sorts of graphical means of calculation that were advanced in this period. We saw various graphical techniques in Chapter 2 and Bjerknes's graphical calculus in Chapter 5. Another example is a graphical procedure that Bjerknes and Sandström presented in 1910 for calculating the heights at which atmospheric pressure equalled certain standard values from tracings of temperature, pressure, and humidity made during a balloon ascent. They write, "The interval of time from the moment the meteorologist has obtained the meteorogram . . . on his desk until he has found [the heights] . . . to be telegraphed ought not to exceed ten to fifteen minutes" (Bjerknes and Sandström, 1910, p. 85).

Development of the graphical calculus continued in this period and was cited in 1933 by the American meteorologist Eric R. Miller as evidence "that meteorology is approaching the standard set by astronomy" (p. 193).[16] Felix Exner contributed to these methods, showing, for example, that the rate of change of pressure is inversely proportional to the area enclosed between consecutive isobars and isotherms (Gold, 1930, p. 196). Solot's article, just referred to as presenting computing forms, also presented a graphical procedure for doing the same calculation, which was finding the depth of precipitable water in a column of air. In 1923 the Japanese meteorologist S. Fujiwhara proposed that forecasters calculate vorticity by using a weather map and a celluloid scale for gradient wind (p. 117). Jerome Namias, who was at MIT in the 1930s, later commented on the interest then shown in graphical methods of solving equations (Namias, 1986, p. 6). Other examples of graphical techniques are the tephigram (discussed below) and many of the procedures used in the search for weather cycles (see Table II). Of all the graphical methods, nomography was most used.

In the 1790s the French government converted its weights and measures to the metric system. To reduce the amount of labor required for this conversion the government published graphic scales with which the conversions could be done quickly. In the course of the 19th century much more elaborate graphical procedures were developed, especially by those doing engineering, ballistics, and meteorology. In 1891 the French engineer Maurice d'Ocagne introduced the word "nomograph,"[17] to denote a figure presenting a quantitative law in such a way that the implication of the law, in any particular case, is readily determinable (usually by seeing where a straightedge, placed so that it connects points on two scales, cuts a third scale). d'Ocagne did much to make nomography a versatile tool (Evesham, 1986, pp. 324–331).

Nomography was widely used in engineering and in many of the sciences. Table III is an incomplete list of major presentations of nomography that were published in the interwar period. The literature of nomography goes back at least as far as the 1840s, when several important works of Leon Lalanne appeared, and continues in the 1980s. Although nomography was still considered a useful discipline in the 1960s, it seems that it has gone largely out of use in the past two decades.[18]

Nomographs were regularly used in meteorology in the interwar period. Figure 1 is an example taken from W.J. Humphreys' *Physics of the Air* (1920). It is a graphical representation of the general gradient-velocity equation:

$$2 \omega v \sin \phi - (1/\rho)(dp/dn) = \pm v^2/r,$$

where the upper sign is used for anticyclones and the lower for cyclones.[19]

To calculate the gradient-wind velocity, place a straightedge so that it passes through the known value of the pressure gradient (on the scale at the left) and the latitude of the place in question (on the scale at the right). Find the intersection of the straightedge and the curve labeled with the known radius of curvature of the local isobar. The vertical line through this point of intersection gives the wind speed.

Brunt and Douglas, in a 1928 paper, derived an equation for estimating the deviation of the observed wind from the wind as calculated by a method similar to the preceding. When calculations were thus attached to calculations, calculating aids became even more important.

The use of graphical techniques by meteorologists in the interwar period was part of a larger movement to substitute analog procedures—slide rules, nomographs and other graphical techniques, and analog machines—for the (digital) computations formerly done with tables. The resulting loss in precision seemed relatively unimportant. And the great gain in speed was everywhere praised. Slide rules were particularly expeditious.

The standard slide rule, which dates back to about 1620, was regularly used by meteorologists. Nelson Haas, in a 1924 article on "A method for locating the decimal point in slide rule computation," stated that slide rules were used extensively in the work of the Weather Bureau and asserted their adequacy:

> Twenty-inch slide rules are used chiefly for such computations. The slide rule is particularly well adapted to this work, for the 20-inch rule yields three figures accurately and the fourth approximately. Four figures represent the maximum accuracy that is readily attainable in meteorological observations, and consequently the 20-inch rule is entirely satisfactory for this work, and it is very expeditious. (p. 29)[20]

Characteristic of meteorology in the interwar period and the succeeding two decades was the development of a great many special-purpose slide rules. Such

Table III

Some of the Major Presentations of Nomography in the Interwar Period

Date	Author	Title	Publisher
1918	J. Lipka	*Graphical and Mechanical Computation*	New York
1918–1920	P. Luckey	*Einfuehrung in die Nomographie*	Teubner, Leipzig
1920	S. Brodetzky	*A First Course in Nomography*	Pitman, London
1920	P. M. d'Ocagne	*Principes usuels de nomographie*	Gauthier-Villars, Paris
1921	R. Soreau	*Nomographie ou traite des abaques*	Chiron, Paris
1921	P. Werkmeister	*Pracktisches Zahlrechnen*	de Gruyter, Berlin
1923	S. Brodetzky	"Nomography," *A Dictionary of Applied Physics,* Vol. 3, pp. 635–644	
1923	B. M. Konorski	*Die Grundlagen der Nomographie*	Springer, Berlin
1923	O. Lacmann	*Die Herstellung gezeichneter Rechentafeln, ein Lehrbuch der Nomographie*	Springer, Berlin
1923	P. Werkmeister	*Das Entwerfen von graphischen Rechentafeln (Nomographie)*	Springer, Berlin
1924	H. Schwerdt	*Lehrbuch der Nomographie*	Berlin
1925	P. M. d'Ocagne	*Esquisse d'ensemble de la nomographie*	Gauthier-Villars, Paris
1926	J. C. Almack and W. G. Carr	*The Principles of the Nomograph in Education*	Bloomington, Illinois
1927	P. Luckey	*Nomographie,* second edition	Teubner, Leipzig
1928	F. Willers	*Methoden der praktischen Analysis*	Berlin
1932	H. J. Allcock and J. R. Jones	*The Nomogram, the Theory and Practical Construction of Computation Charts*	Pitman, London
1936	H. Arkin and R. R. Colton	*Graphs, How to Make and Use Them*	Harper, New York
1937	A. S. Levens	*Alignment Charts*	Wiley, New York
1937	M. G. van Voorhis	*How to Make Alignment Charts*	New York
1939	M. Gorodskii	*Uchenie Zapiski, M.G.U.,* No. 28	

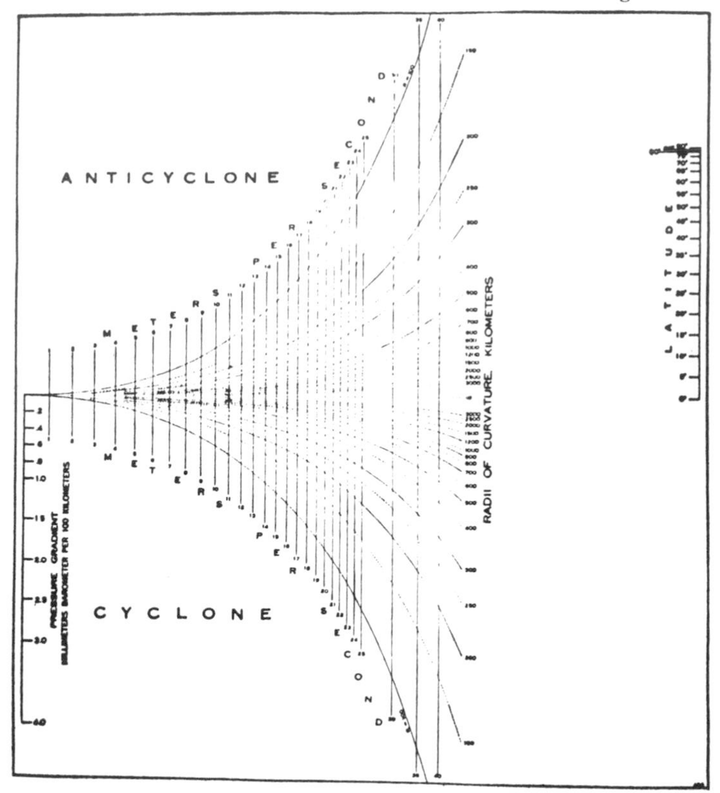

Figure 1 A nomogram, constructed by Herbert Bell of the University of Chicago, for calculating the gradient-wind velocity given the pressure gradient, the latitude, and the radius of curvature of the local isobar (Humphreys, 1920, p. 144).

devices helped forecasters calculate wind direction and speed at various heights. This information could be gained by the use of pilot balloons, but since the raw data were the azimuthal and elevation angles measured by a theodolite much computation was involved. Since using tables was fairly time-consuming, special slide rules were constructed for this purpose and were by 1922 commonly employed (Thompson, 1922, p. 766). Another example is the slide rule, devised by Noel Sellick (1937, p. 439), that allows one to convert barometric pressure at one height to pressure at any other height, for values of the temperature between 0° and 30° C.

Figure 2 shows a hybrid device, a cross between a slide rule and a diagram for graphical calculation. According to its inventor, Leslie Gray, "The device was used to permit rapid computation of more than 350,000 dew points in connection with fire weather data summarizations . . . which otherwise would have been computed laboriously from tables" (1935, p. 16).

A final example is shown in Figure 3. This is only part of an elaborate slide-rule system, designed by J. C. Bellamy in 1943, for calculating the constant

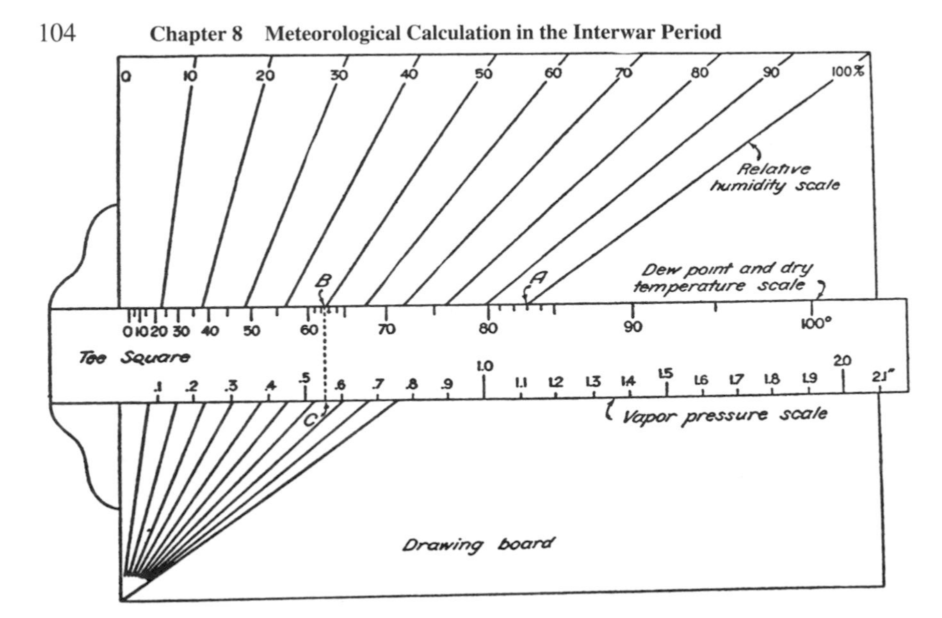

Figure 2 A T-square for hygrometric conversion (Gray, 1935, p. 17).

absolute-vorticity path of wind. The part shown consists of three slide rules with four, eight, and six scales, respectively. The slide-rule system is based directly on Rossby's vorticity equation, which is discussed below.

Calculation in Weather Forecasting

We have seen that a great deal of initial processing of data, such as in converting units of measure or in reducing observed barometric pressure to the corresponding sea-level pressure, was required of forecasters. But once past this stage, once the information was placed on maps, there was little calculation. In a 1941 official publication C. L. Mitchell and Harry Wexler of the Weather Bureau told how a daily forecast was made. The main activity of the forecaster, as they explained it, was entering data on maps:

> Returning to the job of finishing the principal synoptic chart, the forecaster usually is compelled to call a halt on the translation and entry of late land and vessel reports in order to spend a few minutes in computations and in correlation of the conclusions (sometimes contradictory) reached from his brief study of the several charts and cross-sections, before beginning dictation of the forecasts at or shortly after 9:30 a.m. (p. 597)

That is, the forecaster spends "in computations and in correlation of the conclusions" *a few minutes*![21]

The few computations that were performed were often based on formal extrapolation of weather patterns, not on physical theory. Such quantitative, but not

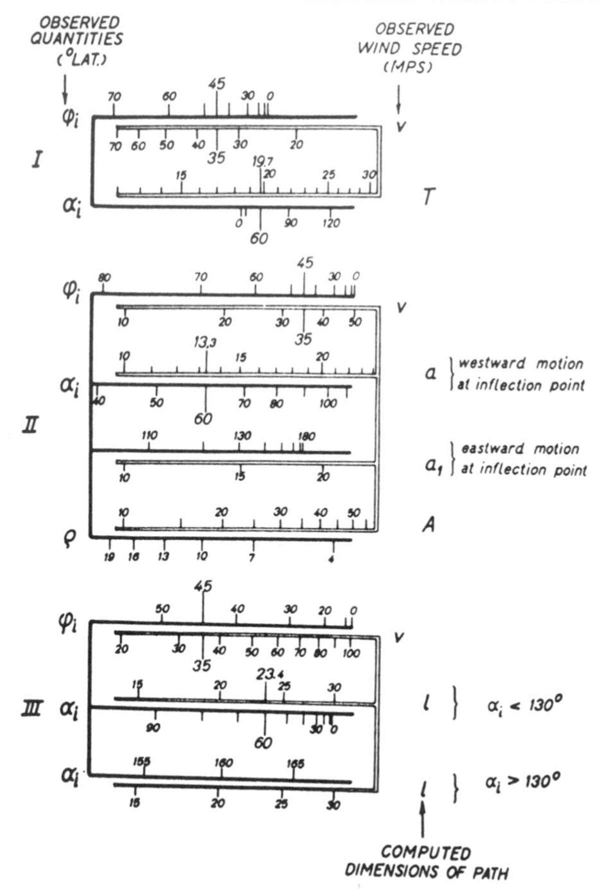

Figure 3 Part of a slide-rule system for calculating the constant absolute-vorticity path (© American Meteorological Society. Reproduced in Godske *et al.*, 1957, p. 716).

physics-based, procedures, such as for predicting the motion of a cold front, began to be used in the years around 1930. G. Dedebant (1927), J. M. Angervo (1928), Antonio Giao (1929), and H. Wagemann (1932) made important contributions to this "kinematical forecasting." The approach was taken farthest by Sverre Petterssen, a Norwegian meteorologist working in the United States, and made widely known by Petterssen's *Weather Analysis and Forecasting, a Textbook on Synoptic Meteorology,* which was first published in 1940. An instant best-seller, the book was translated into more than 20 languages and did much to make forecasting more quantitative (Johannessen, 1975, p. 892).

Only part of Petterssen's book dealt with kinematical rules. Most of it was given to physical theory and to a number of quantitative techniques based on physics. By the late 1930s forecasters were using a number of such techniques. Shaw's tephigram[22] was one of the most important of them. Devised by Shaw in 1925,

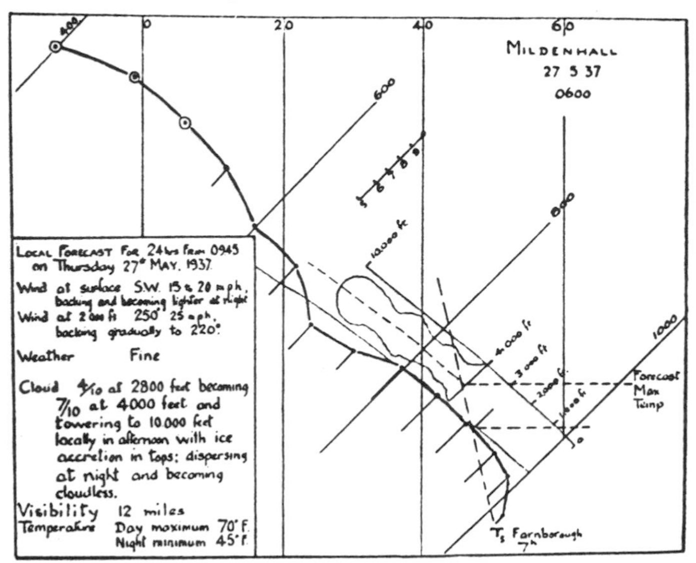

Figure 4 The use of the tephigram to predict cloud formation (Poulter, 1938, p. 280). ("$\frac{4}{10}$ at 2800 feet" means a 40% cloud cover at a height of 2800 ft.)

the tephigram is a diagram showing possible values of entropy and temperature. By plotting on this diagram the line corresponding to a vertical ascent of the atmosphere a meteorologist can predict cloud formation.[23] The use of the tephigram is illustrated in Figure 4.

R. M. Poulter, the author of the article from which the diagram is taken, remarked

> For this work [of cloud forecasting] the tephigram, after a little use, becomes the most reliable and perhaps the most used tool of the forecaster who wishes to make precise weather forecasts, and the close numerical connexion between temperature and humidity on the one hand and cloud height, thickness and amount on the other appears to the uninitiated little short of miraculous, and to the initiated a delightful technical aid. Given an understanding of all a tephigram means, on many days the cloud can be prescribed with something like the accuracy of reading off a logarithm from a book of tables. (1938, p. 278)

In order to use the tephigram in this way one needed data that were not usually available (values of temperature and humidity at many different heights) and an estimate of the daily temperature rise (for which purpose Poulter used climatological tables).

Besides the formal extrapolation of Petterssen and the physics-based diagrams like the tephigram, there was a third quantitative approach to forecasting: the statistical approach. Despite a long and close connection between statistics and meteorology—seen in the works of Laplace, Quetelet, Galton, and Schuster—statistical methods were before the interwar period hardly used at all in weather prediction.[24] During the 1920s, 1930s, and 1940s there were many attempts to use statistics in forecasting, and in the late 1940s forecasts were sometimes presented with probabilities attached.

In 1921 the French meteorologist Albert Baldit in *Études élémentaires de météorologie pratique* deplored the meager progress made in statistical meteorology, and this book was a call for the systematic application of statistics to weather forecasting. The English meteorologist Ernest Gold wrote that the book broke new ground and would be of great utility to meteorologists (Gold, 1923, p. 65). Yet few meteorologists in the 1920s and 1930s tried to use statistics in forecasting. Perhaps the main reason was that statistical forecasting could not compete with a far simpler technique—predicting no change. Baldit thought it would be useful to know the probability of rain between 3 and 4 p.m. on a July afternoon in Paris. This is useful information for a climatologist, but not, as it turns out, for a forecaster. It was not until the late 1940s that statistical forecasting became popular, partly because most statistical methods were fully objective and objectivity in forecasting was then highly prized (see Chapter 9).

There is yet another role for calculation in forecasting: the quantitative evaluation of forecasts. This became a subject of greater interest in the interwar period, partly because meteorologists were learning statistical techniques, which could, of course, be applied to forecasts as well as to meteorological data.[25] An example of this interest is a 1921 article by G. M. B. Dobson in which he proposes a definition of "the improvement due to forecasting."[26] With the great interest in "objective," that is, algorithmic, forecasting methods in the late 1940s (discussed in Chapter 9), the evaluation of forecasts became a field busy with activity and contention.

The Beginnings of Numerical Experimentation

Many meteorologists were primarily interested in understanding atmospheric phenomena in terms of physics rather than in analyzing data or in making forecasts. These meteorologists usually sought to show that some aspect of the atmosphere's structure or action, which had been established observationally, was a consequence of the laws of physics.[27] The demonstration, although mathematical, was seldom strictly deductive and led usually to a qualitative result only.

Occasionally such a demonstration convinced meteorologists of the correctness of data. The use of sounding balloons in the years around 1900 gave evidence that above 10 km or so there was little change of temperature with height. It had long been known that at lower levels the temperature fell off rather uniformly with height. The new evidence was more readily believed after Ernest Gold's 1909 demonstration that such a change of temperature regime is a direct consequence of the laws of radiation and some reasonable assumptions about the content of water vapor and carbon dioxide in the upper atmosphere (Brunt, 1951, p. 120).[28]

Gold's demonstration was quantitative and involved much calculation, but it is clearly distinguishable from numerical experimentation, which is the carrying out of an algorithm (specified beforehand) with the intention of learning something about the physical world or of testing ideas embodied in the algorithm.[29] The usefulness of numerical experimentation to a science depends on how easily algorithms can be carried out, hence on what calculating aids are available.

In the late 1920s G. C. Simpson made a study by numerical experimentation of the absorption and emission of electromagnetic radiation in the earth's atmosphere. The results he reports in three papers published in 1928 and 1929 (Simpson, 1928a,b, 1929). Simpson begins by asserting that the difficulty is one of deducing the consequences of known laws: "No branch of atmospheric physics is more difficult than that dealing with radiation. This is not because we do not know the laws of radiation, but because of the difficulty of applying them to gases" (Simpson, 1928a, p. 70). Saying that "there is no hope of getting an exact solution," he makes a number of simplifying assumptions, which are listed at the outset, so that one can calculate the radiation emitted by the atmosphere.

Simpson makes use of the following devices to facilitate the calculation: the construction of tables for the evaluation of complicated functions,[30] graphical techniques (especially for integrating functions), and computing forms of the Richardson type. It is apparent in all three papers that Simpson is constrained by the labor requirements of certain calculations.

The first numerical experiment gave a surprising result, which Simpson nonetheless affirmed:

> Thus, while admitting that the numerical values may need adjustment as more data become available, there can be no doubt that the general conclusion that the outgoing radiation is almost independent of the surface temperature, and is practically the same in all latitudes, will continue to be true. (1928a, p. 94)

He then conducted a second numerical experiment in which one of the assumptions of the first experiment, that the absorption coefficient of water vapor is independent of wavelength, was replaced by a more realistic assumption, but in such a way that the calculation could still be carried out. This led to quite different results: "The lesson to be learnt from this work is that totally misleading results follow from the assumption that water vapour absorbs like a grey body, and that even qualitative results cannot be obtained on that assumption" (Simpson, 1928b, p. 25). This lesson deserved, according to Simpson, more general notice: "Many problems of atmospheric radiation have apparently been solved by the use of this assumption, and in all these cases the problems must be re-examined using the known absorption of water vapour in the various wavelengths" (Simpson, 1928b, p. 25).

Thus the first two numerical experiments had the effect of testing certain assumptions about the atmosphere. More precisely, they had the effect of showing the sensitivity of the model to assumptions about how water vapor absorbs radiation. Numerical experiments sometimes had the effect of showing the insensitivity of a model to certain assumptions. Since the assumptions depend on theoretical understanding or observational data, numerical experimentation can have the effect of redirecting theoretical and observational efforts. That Richardson's scheme had this effect was shown in Chapter 6. Simpson's work too had this effect, stimulating the careful measurement of the absorption coefficients of water vapor, notably by Walter Elsasser (Rossby, 1941, p. 694).[31]

In the third paper Simpson assumed that the algorithm used in the second paper was reliable.[32] He then applied it to restricted places and times. Thus in the third paper the purpose of numerical experimentation was to learn about the atmosphere.

There are many other examples of numerical experimentation in the interwar period. Simpson's work spawned a school of similar studies, notably by Walter Elsasser, H. G. Houghton, F. Baur, and H. Philipps. Harold Jeffreys did careful calculations of the momentum balance in the atmosphere. David Brunt and C. K. M. Douglas (1928) did calculations on the effect of pressure distribution on rainfall. But the labor required to carry out realistic calculations was too great for numerical experimentation to become an important methodology. Evidence for this view is provided in a paper by R. G. K. Lempfert. He had obtained a surprising result when he had calculated the drying power of the air for a 1-week period. He wondered whether this result was typical and wrote that if a table of drying power, as a function of temperature and relative humidity, had been computed, he could have answered the question (Lempfert, 1932, p. 102).

One of the effects of numerical experimentation is to make precision calculation more important. Since meteorological observations were rarely accurate to more than three or four places, analog devices, such as slide rules and graphical procedures, which were accurate to two or three places, were usually adequate. But they were not adequate when the result depended on the small difference between two large quantities, as it did in some of Margules's and Richardson's calculations, or when long chains of calculations were performed, since errors could then accumulate. The practice of numerical experimentation thus predisposed meteorologists to favor digital over analog calculation.[33]

The larger role for calculation and the use of calculating aids had the effect of making meteorologists more interested in algorithms. Meteorologists were as a result motivated to act as numerical analysts—that is, to study and compare numerical algorithms—and those taking a statistical approach were motivated to act as theoretical statisticians, simply because appropriate statistical tests had in many cases not been invented. Chapter 6 discussed Richardson's work in numerical analysis. Meteorologists who devised statistical tests included J. I. Craig, of the Egyptian Weather Service, who was motivated by his study of winds to extend the statistical method of correlation to vector quantities, and C. E. P. Brooks of the Meteorological Office, who developed many statistical techniques for dealing with meteorological data (Shaw, 1926, p. 270; Rigby, 1958, p. 40).

The English meteorologist David Brunt became both numerical analyst and theoretical statistician. An example of his work in the former capacity is the devising of a criterion for whether to calculate further terms of the Fourier series (Brunt, 1917, pp. 177–179). Also, in a 1928 paper he wrote in collaboration with C. K. M. Douglas, an infinite-series expression for an important quantity is derived, and when it is found that the series "does not appear to converge rapidly" an alternative expression is found. His involvement with statistics was still greater. He invented many techniques for dealing with data, and in 1917 he presented some of these to a wider audience in his book *The Combination of Observations.*[34]

What we have seen then is an increasing role for calculation in the two decades following World War I. The preceding chapter described the steady growth of meteorology, as a science and a profession, in the same period. The following chapter will show that World War II gave a great impetus to both of these trends.

Chapter 9 | The Effect of World War II on Meteorology

Operation Overlord

On the first of June 1944 the weather over northwestern Europe became unsettled. A complex evolving system of low-pressure areas was approaching from the west, and high seas were reported off Ireland and Scotland. To the meteorologists at the headquarters of Operation Overlord in southern England the change was cause for much concern. Overlord, the Allies' planned invasion of Normandy, involving a fleet of more than 5000 ships and landing craft—the largest ever—an air fleet of about 11,000 aircraft—also the largest ever—and some 2 million soldiers, sailors, and airmen, was already under way. On May 29th General Dwight D. Eisenhower, the Allied commander, had set this enormous force into motion, following an extremely elaborate set of plans, to invade on the morning of June 5th. That date had been chosen for the expected moonlight for night-time parachute and glider landings and for the exceptionally low tide just after dawn for removing the underwater obstacles which Field Marshall Erwin Rommel, who was in charge of German defences in France, had ordered built to prevent landings on the beaches.

The success of the operation, however, depended very much on suitable weather. The commanders of the airborne units had stipulated that there be no fog or low clouds, no more than 60% cloud cover, and winds below 20 mph. Each type of air support—high-level bombers, low-level bombers, interceptors, and fighters—had its own weather requirements. The naval bombardment could hardly proceed in high seas or if visibility were less than 3 miles. And, most importantly, for the amphibious landings there could be no fog and on-shore winds could not exceed 12 mph.

The meteorological operations supporting Operation Overlord were complicated. Chief Meteorological Officer was the Scottish meteorologist John M. Stagg, Group Captain of the Royal Air Force.[1] Stagg, who had only a small staff at Overlord headquarters at Teddington, was in continual contact with three major forecasting centers: that of the British Meteorological Office at Dunstable, that of the U.S. Army Air Corps at Widewing, and that of the British Admiralty in London.[2] In addition, Stagg was in continual contact with the staff meteorological officers of the Air Commander-in-Chief of Overlord (Air Chief Marshal Leigh Mallory) and of the Naval Commander-in-Chief (Admiral Sir Bertram Ramsay), in Stanmore and Portsmouth, respectively. Facilities for telephone conferences allowed all six groups to participate in the frequent discussions called by Stagg.

So important were the meteorological considerations that, from June 2nd until the final decision was made, Eisenhower and his commanders-in-chief met twice a day for the sole purpose of hearing the weather forecast and discussing its implications. Before each of these meetings, Stagg conducted a discussion, usually lasting 1 or 2 hours, with the six groups of meteorologists. Since forecasting at that time was a highly subjective process, trying to achieve consensus was an exasperating and only rarely successful endeavor. Making Stagg's task much more difficult was the fact that Eisenhower expected him to give 5-day forecasts, even though Stagg and many other meteorologists doubted that it was possible, except in unusual circumstances, to predict English weather more than 1 or 2 days in advance.

The final decision about whether to attack on June 5th was made at 4:30 in the morning of June 4th, just before the main invasion forces were to leave England. On the basis of the information provided by Stagg, Eisenhower ordered a 24-hour postponement. Many units had already set sail, so this decision necessitated dangerous maneuvering and refueling in stormy seas. Such difficulties, together with considerations of the change in the time of low tide, of the effect of delay on troop morale, and of the possibility that the Germans would detect the invasion force, made it doubtful that another 24-hour postponement could be ordered.

The circumstances in which the final decision was made show the faith the Allied commanders placed in the meteorologists. Eisenhower later described the early morning of June 5th:

> At three-thirty the next morning our little camp was shaking and shuddering under a wind of almost hurricane proportions and the accompanying rain seemed to be traveling in horizontal streaks. The mile-long trip through muddy roads to the naval headquarters was anything but a cheerful one, since it seemed impossible that in such conditions there was any reason for even discussing the situation. (1948, p. 250)

Stagg surprised the assembled officers by reporting that there would be relative calm and good visibility on June 6th, followed by several days of quite variable weather. Aware of the danger that bad weather might leave the first attacking forces cut off from Allied support, Eisenhower decided nevertheless to go ahead with the invasion.

It is probable that Eisenhower thought that the bad weather would make the attack more of a surprise. If so, he was right. As it turned out, German reconnaissance, both by boat and by plane, had been blocked by the storm in progress. Much more important was the fact that German meteorologists had reported an invasion unlikely because of the weather and apparently did not foresee the interval of acceptable weather on June 6th.[3] For the first time in a month German units in northern France were not on alert, and most senior German commanders took the occasion to attend to matters away from their posts—especially significant was the absence of Rommel.

The invasion was carried out successfully. There were of course a great many things that went wrong, but only a few of them were due to the weather (notably the limited visibility that caused many of the airborne units to miss their targets and the seas that proved too rough for the swimming tanks). The information Eisenhower got from Stagg proved reliable: the forecasts for both June 5th (when the weather was completely unfavorable) and June 6th (when it was acceptable) were quite accurate. The Overlord forecasts were regarded as a triumph for the Allied meteorologists, and their importance for the liberation of France as widely commented on.[4]

The Wartime Importance of Meteorology

As we saw in Chapter 7, it was World War I that made a meteorological staff a standard element in military organization. World War II greatly heightened the importance military leaders accorded meteorology. The impact of weather on modern military operations was shown repeatedly in the latter conflict. Major German offensives were slowed dramatically by the early Russian winter of 1941/1942, the most severe on record, and by the sudden spring thaw in the Caucasus in March 1943. The Allies suffered from the fog and dense cloud of the Battle of the Bulge[5] and from typhoons that struck a fleet off the Philippines in December 1944 and a fleet off Okinawa in June 1945. War correspondent Ernie Pyle, when he was with the Allied forces of the Anzio Beachhead in March 1944, reported, "One day of bad weather actually harms us more than a month of German shelling" (Nichols, 1986, p. 243). Eisenhower, in his account of the war on the western front, *Crusade in Europe*, had occasion to comment on the weather about 40 times.

That meteorological information was perceived as having great military value during World War II was shown by the magnitude of the efforts made to gain such information. Besides stationary and mobile weather stations on land,[6] weather-reconnaissance aircraft and weather ships were used regularly by the Allies and by the Germans. The Royal Air Force, which by the end of the war had six and a half squadrons specifically for meteorological flying, made twice-daily flights (each 700 miles out and 700 miles back) to the southwest, to the west, and to the northwest of the British Isles (Dean, 1979, p. 171; Poulter, 1945, p. 391). The U.S. Army Air Corps also had weather squadrons (Hughes, 1970, pp. 82, 85). And as soon as weather information from the west became unavailable, the German weather service (Reichswetterdienst) began a major operation of gathering data over the North Atlantic by weather airplanes involving at least five daily flights; despite heavy losses this operation was continued into 1945 (Mügge, 1948, p. 177). In both the Atlantic and the Pacific, the Allies maintained "high-sea weather stations"—at the end of the war, 22 in the Atlantic and 24 in the Pacific (Hughes, 1970, p. 103; Whitnah, 1961, p. 207).[7] The Germans used submarines to place automatic weather stations in the Atlantic and made repeated expeditions to the northeast coast of Greenland to maintain a manned weather station there.[8] The Allies too gathered meteorological information behind enemy lines, by aircraft, by spies, and by commando units.[9] And the English, Americans, and Germans greatly increased efforts to gather upper-air data by means of radiosonde.

The secrecy of weather information during the war also showed that it was perceived as militarily important. Every country at war stopped releasing such information (Johnson, 1943, p. 200). In England public forecasting was strictly controlled, and no references were permitted to certain types of weather such as gales, snow, fog, and severe frost.[10] In the United States the publication of weather maps was prohibited until the data were at least 1 week old, and public forecasts were intentionally vague.[11] Both by the Allies and the Germans, weather information was, as a matter of course, encoded before being transmitted by telephone or radio. Indeed, this fact played a part in the cracking of the German code Enigma by the English: it was from a German weather ship, captured off Iceland on 7 May 1941, that the English obtained some important documents on Enigma.

It was not only current weather information and weather forecasts that were important to the military: for the first time military leaders made much use of records of past weather.[12] The fact that the war was protracted and geographically dispersed made climatological considerations important in many decisions, such as where naval and air bases should be built, which transport routes should be chosen, what kinds of clothing should be issued to soldiers in different regions, and what heating, air-conditioning, and snow-removal equipment would be needed. Weather records were used to set specifications for buildings, for landing mats, for motor-vehicle lubricants, for transmission lines, and for food-storage facilities (U.S. Department of the Air Force, Air Weather Service, 1949, p. 2). In the long-range planning of offensive operations, commanders made much use of such information; knowing, for example, the probability of fog or overcast skies or high winds over a particular city at a particular time of year was valuable in the planning of strategic bombing. Even in short-range planning, climatological information was important because the military meteorologists often relied heavily upon it in making forecasts (National Oceanic and Atmospheric Administration, 1970, p. 4). Indeed, it was in this period that the phrase "applied climatology" first became common.

It is therefore not surprising that climatological information was also censored, nor that computation of weather statistics became an important activity. The military value of weather records was the reason the British Meteorological Office, at the beginning of the war, stopped publication of the *Daily Weather Report* and the *Monthly Weather Report*; and the U.S. Weather Bureau prepared, for use by the military, weather guides to 26 regions outside the United States, and the Navy's Hydrographic Office prepared another 20 or so [*Q.J.R.Meteorol. Soc.* **66,** 154 (1940); Whitnah, 1961, p. 202; Bates, 1956, pp. 522, 523].[13]

In the United States the wartime need for climatological data had three important consequences: the standardization in methods of observing and recording the weather, the centralization of weather records, and the automation of the processing of weather data. To facilitate the sharing of data, the Weather Bureau, the Navy, and the Army Air Corps agreed to a much higher level of standardization in taking and recording weather observations (National Oceanic and Atmospheric Administration, 1970, p. 4).[14] Before and during World War II weather records were kept in many different places by various civilian and military agencies. The wartime difficulties in using these scattered archives made clear the benefits of a single center for weather records. This was finally achieved in the late 1940s at the New Orleans Port of Embarkation where the Air Force Data Control Unit, the Aerology Section of the Navy, and the Office of Climatology of the Weather Bureau were located (U.S. Department of the Air Force, Air Weather Service, 1949, p. 2).[15]

The Increased Use of Punched-Card Machines

The automation of data processing in meteorology, as in census offices, the insurance business, and elsewhere, was achieved largely by the use of the Hollerith system. As described in Chapter 8, the impossibility of dealing adequately, by hand methods, with the ever-increasing amount of weather data had led to the use of punched-card machines by many national weather services by the mid 1930s. This "data push" became much more intense in the 1940s as the data-gathering accelerated. It forced changes even in the way data were transmitted: meteorologists found it necessary to reduce the size of messages by the use of codes of various kinds (Jenkins, 1945, p. 574).

In the United States the Weather Bureau from 1934 onward had made great use of punched cards for storing and processing data. By the time of the attack on Pearl Harbor, millions of weather observations had been recorded on punched cards, and standard tabulating machines were being used to compute summary statistics and even to print the results in final form automatically. During the war the number of punched cards increased by about an order of magnitude—to about 80 million cards.[16]

More significant than the increase in the number of cards was the increase in complexity of the data processing done automatically. By the end of World War II a wide variety of sophisticated tabulating machines, most of them made by IBM, were being used by meteorologists. There were the simpler machines mentioned in Chapter 8: key punches, verifiers, interpreters, duplicating punches, reproducing punches, and sorters. Two more complex machines that were important to meteorologists during the war were the IBM Type 601 multiplying punch (introduced in 1933) and the IBM Type 077 collator (introduced in 1937).

The basic function of the multiplying punch was to record on an output card the product of two numbers read from input cards; it was capable also of using a common multiplier for a group of multiplicands and of accumulating product totals. The collator compared two numbers, x and y say, on two different cards (from the same or different input decks) and sent these cards to particular output hoppers according to whether x was less than, equal to, or greater than y. Thus the collator allowed automatic arrangement of cards. This machine, like most of the others mentioned above, processed cards at a rate of between 100 and 500 cards/min.

The final, and equally rapid, step in the automatic processing of data was usually carried out by a tabulator that printed the results automatically. These machines (such as the IBM Type 405, introduced in 1934) were quite versatile. A card passed from the input hopper through two reading stations to a particular output hopper. Numbers entered on the card could be printed directly, or compared with a number on the preceding card, or added to or subtracted from a particular register. Control was effected by the information on the card itself (which was read at the first reading station to determine what was to be done with the information on the card at the second station), by controlling switches, and by a plugboard. A simple task for a tabulator would be to produce, from a set of cards containing daily precipitation readings for 1 month at a particular station, such information as the frequency of precipitation, the greatest daily precipitation, and the total precipitation. Since the plugboard was removable, once a plugboard had been prepared for a particular task it would be kept for later use.

Calculator technology developed rapidly in the 1940s. One place this can be seen is in the series of multiplying punches that IBM produced from 1931 into the 1950s: Type 600, 601, 602, 602A, 603, 604, and 605. The first ones, although electrically driven, performed multiplication mechanically. In the 1940s, first relays (electromagnetic switches) and then vacuum tubes (first used in the Type 604, introduced in 1948) were used. In 1949 IBM combined a Type 605 multiplying punch and a Type 407 tabulating machine to produce the so-called card-programmed calculator (CPC). The CPC allowed a whole series of operations, which earlier might have been done in a dozen steps on four or five different machines, to be carried out automatically. In the early 1950s the CPC was being used in the analysis of weather information.

Before the war, the Weather Bureau had used punched-card technology for climatic summarization, such as averages, extremes, and frequency distributions, and in the preparation of several meteorological atlases. During the war the preparation of climatological information remained the main purpose for which punched-card machines were used. Tabulating machines made it much easier to use foreign weather records because of their ability to deal automatically with different units of measurements and different coding practices and formats. This technology made possible the large number of publications, such as the Weather Bureau's *Incidence of Low Ceilings and Low Visibilities in the U.S. Pacific Coastal Regions*, that were prepared for particular military purposes in particular geographic areas. In many instances, punched cards were used directly in making a decision, such as which locations were well suited to an airport, a naval base, or a training center. A leader in the use of punched-card machines for producing useful tabulations of climatic data was Helmut Landsberg, who worked for the Army Air Corps (Mitchell, 1986, p. 257).

The British and the German meteorologists were also making much use of punched cards. The United States captured a deck of about 7 million punched cards prepared by the Germans during the war (U.S. Department of the Air Force, Air Weather Service, 1948, p. 11).[17] The Americans made a copy of this deck—by direct use of a reproducing punch!—and then sent it to the weather services of other countries for their reproducing.

One of the more important uses of automated data processing, developed during the war, was the checking of weather reports for reasonableness and for completeness. Once weather reports were entered on punched cards, a tabulating machine could print a list of missing data, and it could identify many errors by checking whether each observation fell within a predetermined range and by testing for consistency within a single set of observations (such as whether the reported dew point was consistent with the reported temperature and wet-bulb temperature). This permitted the timely identification of errors—which might be due to faulty or miscalibrated instruments, to recording error, or to key-punching error—and hence, in many cases, their rectification.

During World War II punched-card machines were used in new ways in order to improve forecasting. They were used to compute duration frequencies (how often a type of weather continues a given number of hours or days). They speeded the making of weather maps considerably by printing observational data at the appropriate positions on maps,[18] the isolines being then drawn by hand. They were used in correlating surface observations with upper-air information, thus making the latter more valuable. They were used, too, in forecast verification.

One method of forecasting, which made use of punched cards, was peculiar to World War II. The idea behind this method was that records of past weather at two locations, **A** and **B**, and knowledge of the current weather at **A**, could give information about the current weather at **B**, for which no direct information was available. Thus if, for example, a correlation had been found between cloudiness at **B** and wind direction at **A**, then knowledge of the latter could be used in deciding when to undertake a bombing raid on **B**. This method, which was much used during the war, made great data-processing demands—met by the use of punched-card machines—in seeking correlation between weather elements at different locations.

A wartime technique that continued in use after the war, at least in a few places, was "analog selection." The Air Weather Service placed on punched cards information describing each weather map in a 40-year series of maps. The cards were used to find the map of past weather that most closely resembled the current weather map, and the course of the past weather then served as a guide in forecasting. This method was used by the American forecasters at Widewing contributing to the Overlord forecasts (Stagg, 1971, p. 30).[19]

Meteorologists used tabulating machines also for more complex calculations. Here they were following the lead of two great pioneers of scientific calculation,

L. J. Comrie in England and Wallace Eckert in the United States.[20] In the late 1920s Comrie, at the National Almanac Office in London, made calculations of lunar motions. In the 1930s Eckert, at the Thomas J. Watson Astronomical Computing Bureau of Columbia University (the head of IBM having provided the funding), worked to make punched-card machines useful to sciences besides astronomy. He showed, for example, how to use these machines to solve numerically certain classes of ordinary differential equations. Eckert (1940, p. xiv) was effective in proselytizing for his methods, by welcoming visitors to his laboratory, by offering the "Watson Laboratory Three-Week Course in Computing" (which was attended, over a number of years, by 1600 people from 20 countries), and by his 1940 book *Punched Card Methods in Scientific Computation.*

Near the end of the war meteorologists began to use tabulating machines to solve complicated equations. One example is the computation, carried out in 1945 by Gilbert Hunt of the Air Weather Service, of the total amount of water vapor in the atmosphere above a given station; each such computation involved many different operations and hundreds of observations.[21] Another example is a study of the action of winds in the build-up of heavy seas, which involved a lengthy series of operations carried out automatically by a CPC (U.S. Department of the Navy, Aerology Branch, 1953, p. 17).

The use of punched cards by the American weather services continued to grow after the war. By the end of 1947 all Weather Bureau stations were entering, as standard practice, current meteorological observations onto punched cards (Whitnah, 1961, p. 226). In 1953 there were almost 200 million punched cards at the national weather records center, up from 80 million at the end of the war. And, as electronic computers became available to meteorologists in the mid 1950s, the value of having weather information on punched cards became even greater (Bellamy, 1952, pp. 21, 42). Although the principal motivation for all these uses of tabulating machines was to speed up data processing and other computation, a secondary motivation was to eliminate human error in the processes thus automated. Their adoption, then, can be seen as an extension of the century-old tradition of automatic data acquisition, in which the *humanum-est-errare* sense prompted the construction and use of self-recording instruments.[22]

Changes in Meteorological Practice

As a consequence of the military value of current weather information, of forecasts, and of weather statistics, World War II quite suddenly made meteorology a prominent science (Bergeron, 1959, p. 461), and governments[23] therefore made much greater commitments of resources to it. Although on a smaller scale, World War I had had the same effect, so it is not surprising that the two wars are regarded as times of marked scientific progress (Douglas, 1952, p. 1; Waterman, 1952, p. 185). One reason for the wartime progress was that many meteorologists carried out their tasks with a greater sense of urgency and a greater willingness to try new methods.[24] Many other reasons, such as the improvement of observational networks and the stimulus of new demands on the science, are considered below, but perhaps most important was simply the great increase in the number of meteorologists (Smagorinsky, 1972, p. 13).

In the first year of the war there was a sharp increase in the membership in the Royal Meteorological Society, and so great was the demand for forecasting in the following 5 years that the Meteorological Office increased its staff 10-fold (to about 6000) [*Q.J.R. Meteorol. Soc.* **66**, 223 (1940); Sutton, 1955, p. 964]. In the United States some 2000 new meteorologists were trained each year from 1942 through 1945; the strength of the Air Weather Service reached 19,000 officers and men, and that of the Navy's Aerological Service 6000 officers and men (Bergeron, 1959, p. 461; Hughes, 1970, pp. 82, 96).[25] A result of the great demand for meteorological services was that, for the first time, women were employed in large numbers. At the outbreak of the war there were only 2 women in technical positions at the U.S. Weather Bureau; at the end of the war there were 900 (Hughes, 1970, p. 114).

The immediate need for forecasters precluded any lengthy preparation by instruction or apprenticeship. There was no time for acquiring "a sense of the weather" by extensive experience. Exacerbating the problem was the small number of meteorologists available to serve as instructors. In World War I Napier Shaw had coped with the same problem by sending the necessarily inadequately trained officers into the military weather services equipped, in his words, "with a formula by which they could 'carry on' slightly" (in Gold, 1945, p. 227). Something similar was done in World War II.

If a science is presented as a series of algorithms to be learned, then it is easier to train large numbers of people rapidly—and to verify that they have learned something. This is exactly the form of *Workbook in Meteorology*, published in 1942 and written by Athelstan Spilhaus and James E. Miller; according to the authors, "This collection of exercises is the outcome of an attempt to formalize instruction in certain of the elementary phases of meteorology" (p. v). Another wartime textbook in the form of a sequence of exercises is *Weather Principles*, published in 1942 by the Airlines War Training Institute of Washington DC. Even the wartime textbooks in standard discursive format often included many exercises, sometimes with answers—C. G. Halpine's *A Pilot's Meteorology* (1941), David Brunt's *Weather Study* (1942), and the U.S. Navy training manual *Flying the Weather* (1943) are examples[26]—whereas exercises were not commonly a part of earlier textbooks.

This is another force—the attraction of formalized instruction—for the increased use of algorithms. It came to be felt by meteorologists in World War II because of the need to train large numbers of people rapidly, and probably had some slight effect in making meteorology more mathematical.[27] Mathematical procedures became, in many textbooks, the core of the subject, although still accompanied by descriptive material and verbal explanation. Such textbooks influenced what both writer and student thought of as meteorology, and, by rewarding mathematical ability, may have had some effect on the makeup of the next generation of meteorologists.

Although the number of new meteorologists trained each year fell markedly at the end of the war, it nevertheless remained far above the prewar level.[28] In the United States the government employed more meteorologists (both civilian and military) than before, and so did industry (especially the airlines) and universities. The amount budgeted to the Weather Bureau increased from $4.7 million in 1938 to almost $24 million a decade later (inflation accounting for only $6 million of the increase). Some indication of the number of meteorologists employed by the airlines is given by the fact that in a survey (among meteorologists generally) of forecasting practices conducted in 1948, 167 airline meteorologists (from 10 airlines) took part; some indication of the growth of this type of employment is given by the fact that United Air Lines did not have its own weather service before 1937, yet by 1951 maintained three weather centers, staffed by 37 meteorologists (Elliott *et al.*, 1949, pp. 314, 315; Harrison, 1951, p. 106).

The growth of meteorology in academia was even more exuberant. In the few years just before U.S. entry into the war, five American universities established departments of meteorology (MIT, New York University, UCLA, Chicago, and Cal Tech), and until after the war these were the only American schools having autonomous meteorology departments.[29] By 1951, 11 universities had departments of meteorology, 4 others offered graduate training in meteorology, and 82 others offered undergraduate training (Macelwane, 1952, pp. 53–55).[30]

Membership in the American Meteorological Society increased more than threefold in the decade beginning in 1939 (from 1189 members to 3718 members), and the annual expenditures of the Society grew from $5281 to $76,244 (Brooks, 1950, p. 213). The rapid growth during the war and the election of Rossby in 1944 to the presidency led to a major reorganization of the Society in 1945. One of the principal tasks of the reconstituted Society was helping demobilized meteorologists find suitable employment. The Society provided a placement service and worked to stimulate the demand for meteorological services in government and industry.

At the same time the Society became more theoretically oriented. Until then the *Bulletin of the American Meteorological Society* was directed mainly toward practical meteorology and did not include many reports of theoretical research (Byers, 1970, p. 216). This changed after the war. There was, in fact, so much research ready to be published, most of it recently declassified wartime research, that the Society established two new journals: *Journal of Meteorology* and *Meteorological Monographs*.[31] The British and the Germans too had a backlog of unpublished meteorological research at war's end.[32] In the postwar period the amount of research published, like the number of meteorologists, remained far above the prewar level.

A principal reason for the marked scientific progress that occurred in both world wars was the expansion of observational networks. Crucial to the birth of the Bergen School of synoptic meteorology was the 10-fold increase, as a result of World War I, in the number of Norwegian weather stations. One of the most notable effects of World War II was the expansion of the observational network that allowed, for the first time, three-dimensional data analysis as standard practice.

There was, of course, already a long history of gathering upper-air data, which we reviewed in Chapter 5, but it was not until World War II that the collection of upper-air data became abundant and systematic enough to be regularly used in forecasting. Although airplane observations contributed, it was primarily the use of radiosondes—the first successful radiosonde had been launched in 1928—that made this possible. With the coming of war the Germans increased the number of radiosonde stations from about 10 to about 80 (Mügge, 1948, p. 176).[33] In Britain the wartime use of radiosondes increased the amount of upper-air data by one or two orders of magnitude, and the number of radiosonde stations operated by the American civilian and military weather services increased steadily from 6 stations in 1938 to 335 stations (many of them abroad) in 1945 (Poulter, 1945, p. 393; Barger, 1960, p. 18).

In the 1930s German forecasters pioneered in the use of upper-air data, and this work intensified greatly during the war (Douglas, 1952, p. 13; Mügge, 1948, pp. 175–195). In England and the United States, what had been an exceptional practice became standard in daily forecasting. Referring to the British experience, R. M. Poulter wrote

> It may be said as a fairly accurate generalisation that before this war forecasting was done by means of surface weather charts—a two-dimensional problem—but the last six years have seen the training of hundreds of forecasters who now think in three dimensions—north–south, east–west, and ground level to 30,000 feet. (1945, p. 393)

Another result of the new abundance of upper-air data was that physics became more relevant to the forecaster's work and many more calculations came to be performed in everyday forecasting. Besides all the calculations involved in reducing the raw data to the form entered on charts, certain more sophisticated algorithms came to be used because the requisite data were available. Thus the tephigram came to be regularly used, both in Britain and in the United States, to calculate the amount, height, and thickness of cloud (Poulter, 1945, p. 394; Elliott *et al.*, 1949, p. 315).[34] Another example is isentropic analysis. Based on the principle that air moves along surfaces of equal entropy except when there is a gain or loss of heat, isentropic analysis was introduced by Napier Shaw in the 1920s. It was not, however, developed into a practical tool of forecasting until the late 1930s, largely through work directed by Carl-Gustaf Rossby at MIT (1941, p. 653).

It was also Rossby who in 1939 and 1940 introduced two physics-based algorithms (described in Chapter 7): calculation of the speed of a Rossby wave and calculation of a constant-vorticity trajectory. As upper-air data became abundant, many forecasters came to use these algorithms regularly. Determination of constant-vorticity trajectories involved so much calculation that meteorologists devised calculating aids specifically for this purpose: in 1943 a system of seven specially designed slide rules (part of the system is pictured in Figure 3 of Chapter 8) by J. C. Bellamy, in 1945 a set of tables by S. Hess and S. Fomenko, and in 1951 a mechanical differential analyzer by H. Wobus (Godsky *et al.*, 1957, pp. 715–719).[35] As with the tables discussed in Chapter 2, these calculating aids were probably devised not only for saving time, but also with an eye to the *horror mathematicae* of some meteorologists.

The observational networks grew horizontally as well as vertically, and at each station many more measurements were made. There resulted what Jule Charney called "a near discontinuous change" in data gathering (1960, p. 13). Since much of the information was rapidly and widely communicated, forecasters everywhere became overwhelmed with data and hence subject, like climatologists since the late 19th century, to the force of "data push."

This is seen vividly in the contrast, depicted in a 1947 lecture by Ernest Gold, between the English forecaster in 1913 and his counterpart in 1947. The former could deal with all relevant information, coming by telegraph from about 30 stations, as it arrived, and "he could, singly, have manipulated more: he could have brought more within the compass of his single mind." The forecaster of 1947, on the other hand, had to deal with reports from more than a thousand stations, and each report was much more detailed than those issued by weather stations 34 years earlier. Moreover, the forecaster in 1947 made regular use of some laboriously prepared charts—such as prebaratics and 500-millibar charts—which were not used earlier (see Poulter, 1945). The result was that the forecaster needed "a small army of manipulative subordinates" (Gold, 1947, p. 160).

Thus forecasting became a group activity, with specific tasks—such as the preparation of a particular upper-air chart—assigned to each individual (Douglas, 1952, p. 17). This produced new motivations for systematizing and automating the data processing and analysis involved in forecasting: to reduce expenses, to reduce the number of errors, and to control precisely how the data are used to generate the charts, tables, or other aids given to the principal forecaster. Thus meteorologists became more than ever predisposed to welcome any technology promising to facilitate calculation or data handling.

One of the most important chapters in the story of the effect of World War II on meteorology concerns the ways meteorologists were able to take advantage of technology developed for other purposes. Meteorologists have, of course, always been great borrowers of technology, but in this period the effects were enormous. Radar provides one example. Before the end of the war meteorologists were using it to locate storms, to map the features of large storms, and to track pilot balloons (called, in this case, rawinsondes). After the war the U.S. Weather Bureau obtained from the military a large number of surplus radar sets; the Weather Bureau converted them to meteorological use and employed them later in a warning network for tornados (Bigler, 1981, p. 159; Hughes, 1970, p. 124; Whitnah, 1961, p. 232).[36] In the late 1940s the acquisition and initial processing of rawinsonde data were completely automated. This involved automatic tracking and automatic computing (to convert the distances and directions of the balloons into wind speed and direction at successive heights) (Spilhaus, 1950, p. 361); here, besides the radar technology, the technology developed during the war for the automatic aiming of antiaircraft guns probably contributed (to the achievement of automatic tracking). In recent decades several forms of radar have become indispensable tools both in forecasting and in research.

The development of high-altitude jet aircraft, which led to the discovery of the jet stream (Saltzman, 1967, p. 588), is another example of wartime technology put to use by meteorologists. So is the development of high-altitude balloons and of short-wave radio (Waterman, 1952, p. 186; Brunt, 1951, p. 122). The development of rocket technology proved useful in the short term, since captured German V-2s were used in investigations of the upper atmosphere, and in the long term, since it led to the orbiting of weather satellites (Hughes, 1970, pp. 124, 125).[37] Even the development of the atomic bomb had implications for meteorology.[38] But there is no doubt that the wartime technology having the greatest impact of meteorology was the development of the electronic computer.[39]

During the war electromechanical and electronic computing technology was developed in many places. At Bell Laboratories George Stibitz directed the building of a series of increasingly powerful machines. At Harvard, a group headed by Howard Aiken built the Mark I, which was largely mechanical, and the Mark II, which was based on the electromagnetic relay. The Code and Cipher School in Bletchley Park, England, was the birthplace of several computers. In Germany Konrad Zuse built several machines, including one (the Z3) that is considered the world's first fully functional computer with automatic control of its operations (Williams, 1985, p. 222). And most important was the development of the Electronic Numerical Integrator (ENIAC) at the Moore School of Electrical Engineering of the University of Pennsylvania. The ENIAC, the world's first large-scale electronic computer, was designed by John Mauchly and Presper Eckert. Mauchly we met in Chapter 8, as a physicist driven by "data push" to devise calculating schemes and calculating devices. Except for having interested Mauchly in computing technology, meteorology does not seem to have played a part in the wartime development of computing technology. In the postwar period, however, the history of computers and the history of meteorology overlap considerably.[40]

Changes in Meteorological Research

A notable development in the 1940s that was in part a result of the newly abundant upper-air data was the appearance of a large number of "bookkeeping" studies. In these studies the total value of some variable in some region (often the atmosphere as a whole) is calculated by quantifying all the processes that increase or decrease the value of the variable. An example is the study of the carbon dioxide content of the atmosphere, or the heat stored in the sea, or the angular momentum of a hurricane. In such studies the geographic distribution of the variable is not usually of interest, only its total value and the ways this value is increased or decreased.

These studies were not new to the 1940s. In a famous paper of 1903 Max Margules studied the energy content of a storm, and in 1920 Richardson published an important paper entitled "The supply of energy from and to atmospheric eddies." In 1926 Harold Jeffreys initiated the study of the momentum balance of the atmosphere (Bolin, 1952, p. 97); G. C. Simpson, 2 years later, completed the first detailed calculation of the radiation absorbed and emitted by the atmosphere.[41] These cases notwithstanding, it was not until the 1940s that bookkeeping studies became a major methodology.

Studies of the momentum balance of the atmosphere became common in the years around 1950; two studies by Jacob Bjerknes and Victor Starr published in 1948 were particularly influential. At about the same time, studies of the energy balance of the atmosphere also became common; James E. Miller, Victor Starr, and J. Van Mieghem did important early work. These studies often involved the statistical use of large amounts of upper-air data, and in many cases the requisite data were not available before the 1940s.[42]

In the early 1950s the variety of bookkeeping studies increased: at MIT the radiation balance of the atmosphere was studied; at New York University, the atmospheric heat balance; at the University of Washington, the heat budget of the upper 15 meters of the earth; and at the U.S. Navy's Hydrographic Office, the ice budget of the Arctic Pack [*BAMS* **42**, 103–104 (1951); Eriksson and Welander, 1956, p. 155; *BAMS* **32**, 108 (1951); *BAMS* **33**, 34 (1952); Bates, 1956, p. 524]. The ozone content and carbon dioxide content of the atmosphere were studied, and in subsequent decades bookkeeping studies of atmospheric constituents (tritium, methane, sulfur, and a great many others) became a standard methodology. These studies often required much calculation. For studying the emission and absorption of radiation in the atmosphere, graphical methods of calculation were developed independently by Walter Elsasser in the United States, by G. D. Robinson in England, and by Fritz Möller in Germany (Brunt, 1951, p. 122; Landon and Raschke, 1983, p. 1093). In recent decades such studies have made much use of computers. Since these studies have been the work of theoretical meteorologists seeking to connect theory with data, "theory pull" may be said to have been operating.

The wars stimulated scientific and practical progress also because they made new demands on meteorologists. In World War I the close cooperation between meteorologist and military leader required a new type of forecast, the detailed short-range forecast, and the use of poison gas led to much research on the diffusion of a gas in different meteorological conditions (Douglas, 1952, p. 3; Brunt, 1951, p. 121). In World War II, when planes flew at much greater heights, forecasters were asked about clouds and ice formation (on a plane's wing or in its carburetor) at these heights. They were asked whether there would be fog when planes returned from night-time bombing. They were asked about high-level winds all along a 500-mile flight path, about local conditions hundreds of miles behind enemy lines, about the advisability of lowering barrage balloons when there was a chance of lightning, and about many other matters. New techniques, and occasionally new scientific understanding, resulted.[43]

Sometimes, though, wartime demands resulted, not in practical or scientific advance, but in the recognition that certain tasks were hardly possible. This was the case with the urgent need for longer-range forecasts. Meteorologists had always been asked to increase the range of their forecasts, and many attempts had been made. Yet until 1940 neither the British nor the U.S. weather services issued forecasts for more than 2 or 3 days in advance, and many, perhaps most, meteorologists thought that predicting beyond 1 or 2 days was not scientifically justifiable.[44]

However, in 1940 the U.S. Weather Bureau began issuing 5-day forecasts. This was the direct result of a research project at MIT, which Rossby had directed in the late 1930s.[45] The accuracy of the Weather Bureau's 5-day forecasts for the period from 1 October 1940 to 30 June 1941 was found to be 48% for temperatures and 16% for precipitation (Whitnah, 1961, p. 166), a result which both the proponents of 5-day forecasting and the skeptics could take as confirming their views.

Ernest Gold, who had been in charge of the British meteorological service in France in World War I, knew that longer-range forecasts would be of immense value militarily, so shortly after the outbreak of World War II he organized a group within the Meteorological Office at Dunstable to investigate the matter. The fact that abundant upper-air data were just then becoming available gave grounds for optimism, as did the Meteorological Office's recruitment in the fall of 1941 of Sverre Petterssen (head of MIT's Department of Meteorology) to direct a group in developing longer-range forecasting by the use of the upper-air data. In the next years, tests of many procedures were made by the meteorologists at Dunstable, but no procedure was found dependable enough to be put into practice.

Nevertheless, soon after assuming command of Operation Overlord in January 1944, Eisenhower made it known that, because Overlord would have to be set in motion several days before the actual invasion, he expected day-to-day forecasts for 5 days ahead, and Chief Meteorological Officer Stagg was ordered to begin producing them. Each Sunday evening from mid February until June, Stagg consulted with the other meteorologists to prepare the Monday-through-Friday forecast. Since the American forecasters at Widewing were much more willing to make longer-range forecasts than were the British, almost all of the details given for the latter part of each 5-day period came from the Americans (Stagg, 1971, p. 32).[46]

In May a meteorologist at Overlord headquarters made an analysis of all their 5-day forecasts, comparing them with the actual weather. It was found that the forecasts were reliable for little more than the first day of each period, and that if the second day's forecast had been simply repeated for the third, fourth, and fifth days, the results would have been no worse (Stagg, 1971, p. 57). Stagg gives no indication that this analysis was communicated to the Overlord commanders. Indeed, there seems to have been general satisfaction with the forecasts, and, in any event, the 5-day forecasting continued.[47]

This incident illustrates two points. The first is the difficulty of improving forecasts. From the late 19th century to the computer era there was little improvement in either the time range or the accuracy of forecasts, this despite continual effort on the part of meteorologists in many countries. Forecasts became, especially after World War I, much more detailed, but for those elements that had always been predicted—such as precipitation and temperature—there seems to have been only slight increases in predictive accuracy.[48] The second point is that the consumers (here, the military leaders) have often been satisfied with forecasts that meteorologists (even the ones issuing the forecasts) have judged to be of little or no value.[49]

In the last half-century the government sponsorship of research has been of immense importance for meteorology. It was during World War II that this practice began on a large scale. In the United States before the war there was little government funding of meteorological research, and until the mid 1930s even the Weather Bureau did little research. The war permanently changed this. Government funding of meteorological research—like government funding of scientific research generally—stayed at a high level after the war. This was due partly to the impressive success of government-sponsored wartime research and partly to the fact that the Cold War followed on the heels of World War II. The funding of scientific research, even that not directly applicable to national defense, was often justified as contributing to national security in the long term. Thus President Truman, in requesting $15 million for the National Science Foundation in the 1953 federal budget, said that "a strong, steady and wide ranging effort in science is as essential to our sustained national security as the production of weapons and the training of military personnel" (Waterman, 1952, p. 184), and meteorology was named as a field of special interest in the Foundation's 1953 budget.

All told, the U.S. government provided about $5 million a year for meteorological research in the late 1940s (Thompson, 1987, p. 632).[50] A member of the Department of Meteorology of New York University in 1951 testified to the change: "Following the general trend more and more of this work has been sponsored by the federal government through its various military and civilian agencies."[51] At MIT, for example, in 1951 the Air Force was supporting three projects, the Weather Bureau, two projects; and the Office of Naval Research and the Army's Signal Corps, one each.[52]

Most of the government-sponsored research—whether carried out by meteorologists employed by government, industry, or universities—was directed toward specific questions. As we have seen, the war placed demands on meteorologists, and these demands raised new questions in climatology, in atmospheric processes (such as cloud formation and aircraft icing), and in forecasting. The demands were largely the same in all combatant countries. Thus, although there was during the war little communication between American and German meteorologists, the research done in the two countries was remarkably similar. Helmut Landsberg, in a

1950 report on Ratje Mügge's *FIAT Review of German Science 1939–1946*, wrote, "One is forcibly driven to the conclusion that at least for a period as long (or as short) as seven years work in different nations proceeds essentially along the same lines; rates of advance and discoveries are generally parallel" (Landsberg, 1950, pp. 67–68). Landsberg goes on to attribute the similarity to the similar military needs in the two countries: "To a considerable extent, the demands of warfare steered the German effort into channels that were similar to those followed by U.S. meteorologists and the results seem to have been comparable."

An important legacy of the war was the large-scale government-sponsored research program aimed at a specific result. In the postwar years there were a number of such programs in meteorology. For example, from 1945 to 1949 Horace Byers directed a major study of thunderstorms; the project was a joint undertaking of the Weather Bureau, the Air Force, the Navy, the National Advisory Committee for Aeronautics, and the University of Chicago. And when this project ended, Byers immediately assembled another team of scientists to study the microphysics of convective clouds. Project Helios and Project Skyhook were large-scale projects (sponsored by the Office of Naval Research) that involved scientists and engineers from industry, universities, and government. Project Cirrus, a 5-year study of the possibilities of cloud modification, began in 1946 with support from the Army, Navy, and Air Force. In 1953 the Great Plains Turbulence Field Program involved scientists from 10 universities and five government agencies (U.S. Air Force, 1957, p. vii). The most important of these projects for the history of meteorology, von Neumann's meteorology project at the Institute for Advanced Study in Princeton, is the subject of Chapter 10.

World War II strengthened the ties between the U.S. Weather Bureau and the military: joint facilities—such as the Weather Bureau–Air Force–Navy Analysis Center (formed in 1947) and the Joint Numerical Weather Prediction Unit (formed in 1954)—and joint research projects became common.[53] The war greatly increased the amount of cooperation between the U.S. Weather Bureau and foreign meteorological services, and it was during the war that the U.S. weather services began a program of worldwide data procurement (Whitnah, 1961, pp. 208–210; Barger, 1960, p. 3).

Interest in Objective Forecasting

We have seen how the war affected observational techniques, forecasting practices, the direction of research, and the institutional setting of meteorology. Another effect of the war was something less tangible: a general interest in objective forecasting. Objective forecasting was defined to be forecasting that did not involve personal judgment, so that "if two forecasters were given copies of one manual describing a forecast method and placed in separate rooms with current data, they would make identical forecasts" (Gringorten, 1953, p. 57). One might say, then, that World War II caused forecasters to feel keenly the attraction of "science-not-art" and to wonder whether forecasting could be made fully algorithmic.

One reason for this interest was that during the war consensus among forecasters was often vital. Whenever military units, each having its own meteorological staff,[54] undertook joint operations, coordination of action required that the respective commanders base their decisions on the same meteorological information. Early in the war the British Bomber Command had found it necessary to ensure that there was agreement between the forecasts issued by different Bomber Group headquarters since many missions involved more than one Bomber Group. To this end the meteorologists at each Bomber Group and those at the Central Forecast Branch held daily telephone conferences, and they usually did manage to reach agreement, although not without heated argument (Johnson, 1943, p. 202; Stagg, 1971, p. 9). The fact that agreement was usually reached may have owed much to the military setting of the discussion.

In making forecasts for Operation Overlord, John Stagg tried to achieve consensus among the six groups of meteorologists partly as a means of reaching the most accurate and likely forecast possible, but primarily to ensure the coordination of meteorological advice to the different units involved. Stagg describes some of the telephone conferences he conducted just before D-Day. There were continual disagreements about whether earlier predictions had been borne out or not, about the analysis of the current weather map, and about future developments. Of the conference on the evening of June 2nd, Stagg says

> Had it not been fraught with such potential tragedy, the whole business was ridiculous. In less than half-an-hour I was expected to present to General Eisenhower an "agreed" forecast for the next five days which covered the time of launching of the greatest military operation ever mounted: no two of the expert participants in the discussion could agree on the likely weather even for the next 24 hours. (1971, p. 86)

This particular conference was unusual in the degree of dissension, but ready agreement was probably no less rare.[55]

The fact that consensus was important made algorithmic forecasting procedures more attractive, since they could be easily communicated and, once adopted, seldom led to disagreement. Of course, consensus might result—as it often does among wine-tasters or music critics—from judgments that are not the result of following a specified procedure. But such consensus usually requires an extended period of shared experience, something very few World War II forecasters had.[56] Moreover, since the subjective procedures for reaching judgment are often inscrutable, when disagreements do arise there is simply an impasse, not unlike the *de gustibus* impasse.

There were frequent impasses in the discussions Stagg describes. According to Stagg (1971, p. 52), the forecasters were as a rule not eager to discuss the processes by which they arrived at a particular forecast. One of the participants, C. K. M. Douglas, later wrote that forecasting procedures were "inherently indescribable" (1952, p. 1). And questioning the judgment of another forecaster—

since that judgment resulted from a complex, not fully conscious process that could not be unscrambled—often led, in Stagg's words, to "argument of a kind encountered only when personal integrity is questioned" (1971, p. 67).

In the U.S. Navy there was perhaps even greater dissatisfaction with the subjectivity of forecasting:[57]

> Probably, the most disconcerting problem to the fleet forecaster [during and subsequent to World War II] was the extreme subjectivity represented in the hand product. It is well known that in sparse data regions, such as the oceanic areas in which fleet operations take place, a group of analysts given identical data will generally produce as many different analyses as there are analysts.

The lack of agreement among forecasters was a major problem during the Berlin Airlift, which lasted from June 1948 until the following May, and also during the Korean War, which began in June 1950 (Bates and Fuller, 1986, pp. 144, 159).

The need for consensus enhanced, therefore, the appeal of "science-not-art." So did the increase in relevant data. There seemed to be a need for fully specified procedures that, as R. C. Sutcliffe and A. G. Forsdyke of the Meteorological Office put it, "will allow the new knowledge to be used *systematically*," because "there is so much the forecaster can take into account that he hardly knows where to begin or what weight to assign different factors" (1950, p. 189, emphasis added). This, of course, is "data push" being felt by the forecaster. Finally, there were two other factors (both discussed earlier): the need to train forecasters quickly and the fact that forecasting had become a group activity. Both seemed to call for fully specified procedures.

All of these factors, added to the perennial attraction of science-not-art, gave rise to a widespread interest in objective forecasting. In his 1946 Presidential Address to the American Meteorological Society H. G. Houghton set as a goal "an objective method of forecasting based entirely on sound physical principles," although he thought that there was no immediate prospect of attaining it (in Douglas, 1952, p. 16). Rossby was expressing the feeling of many meteorologists when he wrote in 1949 of "the horrible subjectivity" of traditional forecasting (in Platzman, 1979, p. 308).

As documented in earlier chapters, the attraction of science-not-art had long been felt: forecasters had always sought explicit rules to provide some guidance in making forecasts. This was even the case with the Bergen meteorologists, whose forecasting procedures were probably *more* subjective than earlier procedures. One of the founders of the Bergen School, Tor Bergeron, wrote

> The endeavour underlying all the work of the Bergen School was to minimize the previous unnecessarily great subjectivity of forecasting by trying to arrive at explicit and physically explainable rules for the displacement and development of well-defined weather systems. . . . (1959, p. 458)

What was new in the late 1940s was the feeling that the entire forecasting process should be fully specified. The terms "objective" and "subjective," which hardly occur in the meteorological literature before then, were applied in an asymmetric way: forecasting that involved any personal judgment was "subjective forecasting," whereas "objective forecasting" was, by definition, fully algorithmic. One can then distinguish between the attraction of science-not-art, which had long been felt, and the interest in objective forecasting, which became widespread only in the late 1940s. It is true that Vilhelm Bjerknes wanted to "calculate" the weather, and that Richardson specified his forecasting scheme in 23 computing forms, but Bjerknes and Richardson were not particularly interested in objectivity *per se*. What they wanted most was to use physics to predict the weather, hence were responding primarily to "theory pull" rather than the attraction of science-not-art.

In any event, it was not until the late 1940s that articles on fully algorithmic forecasting became common. Some articles dealt with general algorithms; examples are John Bellamy (1949), "Objective calculations of divergence, vertical velocity and vorticity"; H. A. Panofsky (1949), "Objective weather map analysis"; and R. C. Sutcliffe and A. G. Forsdyke (1950), "The theory and use of upper air thickness patterns in forecasting." Most of the articles, though, presented forecasting procedures designed for specific locations; examples are Edward Vernon (1947), "An objective method of forecasting precipitation 24–48 hours in advance at San Francisco, California"; Samuel Penn (1948), "An objective method for forecasting precipitation amounts from winter coastal storms for Boston"; Woodrow Dickey (1949), "Estimating the probability of a large fall in temperature at Washington, D.C."; and J.C. Thompson (1950), "A numerical method for forecasting rainfall in the Los Angeles area."

One of the arguments made for objective forecasting was that one could then assign probabilities to forecasts (Gringorten, 1953, p. 58). A subjective forecaster may well have a feeling of the uncertainty of a particular forecast, but to convey to others that feeling of uncertainty, as by assigning a number to it, was problematic. By contrast, for many forecasting algorithms there was a straightforward, although computationally demanding, way to assign a probability: apply the algorithm to past data and compute the percentage success for each type of forecast.

In a 1953 review article Irving Gringorten presents a graph of the number of articles on objective forecasting (reviewed in *Meteorological Abstracts and Bibliography*) as a function of year of publication (p. 59). Until the mid 1940s there were, with only a few exceptions, fewer than three such articles a year. In 1940 a steep rise began: from 1 in 1940 to 10 in 1945 to 42 in 1950. Many of these papers presented a statistically based algorithm, often arrived at by a straightforward regression analysis.

In a 1944 paper in the *Quarterly Journal of the Royal Meteorological Society*, the South African meteorologist T. E. W. Schumann explained this method: "The statistical method essentially amounts to the prediction of the weather element **P** at a station **A** by means of the regression equation

$$\mathbf{P} = \mathbf{k}_1\, \mathbf{x}_1 + \mathbf{k}_2\, \mathbf{x}_2 + \ldots + \mathbf{k}_n\, \mathbf{x}_n$$

in which $\mathbf{x}_1 \ldots\ldots\ldots \mathbf{x}_n$ are the values of a number of weather elements observed at the control stations $\mathbf{B}_1 \ldots\ldots\ldots \mathbf{B}_n$ a specified time before $\mathbf{P}$ is observed, and the regression coefficients $\mathbf{k}_1 \ldots\ldots\ldots \mathbf{k}_n$ are computed from previous records of the weather elements $\mathbf{P}, \mathbf{x}_1 \ldots\ldots\ldots \mathbf{x}_n$." Schumann wrote that the method had received little attention from day-to-day forecasters. Two explanations were offered in the paper and the subsequent discussion: a number of earlier attempts had given disappointing results, and the computational labor was excessive. In 1945 C. F. Sarle, Assistant Director for Scientific Research of the Weather Bureau, said that much more computation for statistical purposes would be done if the cost of computation could be reduced.[58]

Means of facilitating the computation became more readily available in the 1940s. In 1942 Schumann and W. L. Hofmeyer presented a simplified method of computing correlation coefficients.[59] The New York Mathematical Tables Project (mentioned in Chapter 8) was commissioned by the Weather Bureau just after the war to find a statistical description of wind velocities in a certain area (Campbell-Kelly and Williams, 1985, p. 349). Tabulating machines were used as early as 1934 for this purpose, and a few machines were built specifically for computing correlation coefficients (Conrad and Pollak, 1950, pp. 425, 426; Eames and Eames, 1973, p. 89). The facilitation of regression analysis by such means led first, in the years around 1950, to a large number of articles on forecasting algorithms based on regression, and then, in the mid 1950s, to the falling out of favor of this method.[60] In 1954 O. G. Sutton wrote, "The statistical approach has attracted, possibly, more workers than any other; but to-day it is doubtful if anyone (apart from a few enthusiasts) would claim that the results obtained are commensurate with the effort expended *or that the method merits further serious study*" (p. 113, emphasis added). As with the weather cycles discussed in Chapter 8, this disenchantment would probably not have come so abruptly without the new availability of calculating aids.

The disenchantment with statistically based procedures did not put an end to the desire for objective forecasting. In 1953 an advisory committee of eight meteorologists recommended that the U.S. Weather Bureau expand its research program considerably and that the research "be directed toward the development of automatic observing stations with compatible communications, automatic data collecting and filtering systems, *objective machine analysis and forecasting techniques*, and improved high-speed communications" (U.S. Department of Commerce, 1953, p. 37, emphasis added). At an international conference in Rome in 1954 T. E. W. Schumann propounded the following two axioms (among others): "Synoptic forecasting, mainly subjective as it is, holds out no prospects of any fundamental progress," and "The ideal to be striven for is the abolition of synoptic forecasting and its replacement as soon as practicable by objective methods" (in Bergeron, 1959, p. 470).

Although there were a greater number of objective procedures than ever before—use of the tephigram, calculation of constant-vorticity trajectories, and isentropic analysis being the most important additions—subjective procedures were still at the heart of forecasting. The reason for this was simply that no one had found an algorithm that could do as well as the subjective forecasters. In 1952 C. K. M. Douglas said of temperature forecasts, " . . . it is beyond doubt that the most important requirement for a high standard is long experience of forecasting temperature at the place or places which are checked and not an expert knowledge of the latest upper-air techniques." Douglas said also that "A good forecaster is one who is more likely to know when *not* to apply the text-book rules. . . ." (1952, pp. 2, 16). And in 1954 O. G. Sutton reaffirmed that forecasting was an *art* in which good judgment and long practice were essential (p. 1114).

This interest in objective forecasting was important in preparing the soil for the flourishing of numerical forecasting in the 1950s and 1960s. It made the development of numerical algorithms more attractive to researchers. It made forecasters more willing to try the numerical procedures. And it made governments more willing to provide the funding.

Part III | The Beginning of the Computer Era in Meteorology

The scheme is complicated because the atmosphere is complicated. But it has been reduced to a set of computing forms.

LEWIS FRY RICHARDSON, 1922

Effective use of the wealth of climatological data, explanation of observed atmospheric phenomena on the basis of physical principles, and the ability to draw on data and physics to generate timely forecasts—all these made great calculational demands. The electronic computer met these demands and, in doing so, helped bring about the unification of meteorology envisioned by Vilhelm Bjerknes at the beginning of the century.

Chapter 10 John von Neumann's Meteorology Project

In 1922 Lewis Fry Richardson wrote, "Perhaps some day in the dim future it will be possible to advance the computations faster than the weather advances and at a cost less than the saving to mankind due to the information gained. But that is a dream." In 1946 this was no less a dream. The two fundamental difficulties remained: to show that an algorithm, based on physical laws and observational data, can predict the weather more accurately than human forecasters, and to show that such a calculation can be carried out fast enough to give some foreknowledge of the weather. Richardson's monumental attempt had shown how great these difficulties were in 1920, and developments up to 1946 had not improved the prospects much.

One of the topics at a meteorological conference at the University of Chicago in December 1946 was numerical forecasting. The conferees agreed that "the time is not ripe for . . . a numerical forecast system."[1] And at another conference that same month H. G. Houghton, president of the American Meteorological Society, said, "There appears to be no immediate prospect of an objective method of forecasting based entirely on sound physical principles"(in Douglas, 1952, p. 16). Yet less than 9 years later, by the summer of 1955, the U.S. Weather Bureau was using an electronic computer to make daily forecasts that were as rapid and as accurate as those of human forecasters, and meteorological offices in other countries were soon doing the same. One man, a nonmeteorologist, was largely responsible for the rapidity of this change: John von Neumann.

Von Neumann's Interest in Meteorology

Von Neumann, born in Budapest in 1903, moved to Princeton in 1930 and became a permanent member of the newly established Institute for Advanced Study in 1933. He made major contributions to logic, measure theory, the theory of Lie groups, and the theory of Hilbert spaces. But perhaps his most important work was in applied mathematics, where, according to the mathematician Jean Dieudonné (1976, p. 89), he was certainly the equal of Gauss, Cauchy, or Poincaré. Von Neumann achieved fame for his axiomatization of quantum mechanics; the formalism he introduced in the years 1927 to 1932 is still in use. He was the principal founder of a new branch of mathematics, game theory, and he showed its applicability to economics. And he is perhaps best known today for his work with computers, which encompassed abstract theory, logical design, engineering design, programming, numerical analysis, and applications of computing.

Von Neumann apparently became interested in computation while working as a consultant to the Army Ordnance Department at Aberdeen, Maryland. As a result of this work, which began in 1937, he became an expert on theories of shock and detonation (Aspray, 1987, p. 171). He worked for the Manhattan Project at Los Alamos, New Mexico, from 1943 to 1945. There he needed to solve complex problems of hydrodynamics, and, in common with many other engineers and applied mathematicians of the 20th century, he therefore became interested in numerical methods of solution and in the use of calculating aids.

Von Neumann had, however, a second, quite unusual, reason for interest in computation: he thought that it could be a primary means of advancing mathematical and scientific theory. Von Neumann's own wartime work shows the value of heuristic computation; in a 1944 paper, for example, his general analysis of shock waves depended on computations carried out on punched-card equipment. Von Neumann believed that great advances in hydrodynamics had resulted from wind-tunnel experiments and that this experimentation was quite different from most in physics—the purpose was not to establish laws but to perform integrations that could not be done analytically or numerically. He thought that with electronic computers such integrations—and therefore such experimentation—could be carried out numerically.[2]

In 1944 von Neumann became associated with work going on at the Moore School of Engineering at the University of Pennsylvania, namely the building of the ENIAC, the first electronic computer, and the planning of a much more versatile and powerful computer, later named the EDVAC.[3] In June of the following year he wrote *First Draft of a Report on the EDVAC*, which was, according to the historian Michael Williams, the "document which first described, in detail, the concept of the stored program digital computer" (1985, p. 302). By December 1945 von Neumann had succeeded in convincing both the Institute for Advanced Study and Radio Corporation of America, which had a research facility near Princeton, to support an electronic-computer project.[4] Von Neumann's objective was to build a powerful computer to be used to advance the mathematical sciences.

It appears that until December 1945 or even later von Neumann did not consider meteorology as a principal area for application of the proposed computer. In a request for funding that von Neumann sent on 28 November 1945 to the Navy's Office of Research and Invention he came no closer to naming meteorology than in the following passage: "High speed calculation to replace certain experimental procedures in some selected parts of mathematical physics. We think at this time of parts of fluid dynamics, elasticity and plasticity theory, optics, electrodynamics, and quantum theory of atoms and molecules."[5] Yet by May 1946 von Neumann had decided on the forecasting problem as the *main* application of the proposed computer.

How his attention was drawn to meteorology is not clear.[6] Rossby, whom von Neumann had met in 1942, had probably discussed the forecasting problem with him. Reichelderfer, Chief of the Weather Bureau, in looking for more powerful calculating aids, may have contacted von Neumann.[7] In any case, it is likely that the most important influence was Vladimir Zworykin, an electrical engineer at RCA.[8]

Zworykin believed that new computing technology would make possible prediction and control of the weather.[9] By 6 November 1945 his proposal to develop a computer specifically for meteorology was known to Athelstan Spilhaus, chairman of the meteorology department of New York University, for on that date Spilhaus wrote to Reichelderfer about it.[10] On 25 November Reichelderfer wrote to Zworykin, inviting him to Washington. Zworykin suggested that von Neumann, whom Zworykin had met 2 weeks earlier, join the meeting, which took place 9 January 1946 (Aspray, 1990, p. 130).[11] Two days later an article appeared in the *New York Times*, announcing that

> Plans have been presented to the Weather Bureau, the Navy and the Army for development of a new electronic calculator, reported to have astounding potentialities, which, in time, might have a revolutionary effect in solving the mysteries of long-range weather forecasting.
>
> If the super-calculator could be built and operated successfully, weather experts said, it not only might lift the veil from previously undisclosed mysteries connected with the science of weather forecasting, but might even make it possible to "do something about the weather" through advance application of scientific knowledge concerning counter-measures to unfavorable weather. The benefits to agriculture, shipping, air travel and other activities through advance knowledge of the weather were obvious, they added, and the savings in crops, money and even lives would be incalculable.

This newspaper article, one of the more important in the history of science, had a decided effect on two groups of people, the developers of computers and the meteorologists, giving an unpleasant shock to many of the former and a galvanic stimulus to some of the latter. Mauchly, Eckert, and others working to build computers were angered by its implication that von Neumann and Zworykin and others at RCA comprised the leading group in electronic computing and that theirs was the only project of any consequence (Williams, 1985, p. 347)[12]—this at a time when information about ENIAC and EDVAC was still classified.[13] For some meteorologists it suggested that Richardson's dream of computing the weather might become a reality, and it aroused their interest in von Neumann's computer.[14]

Von Neumann's decision to focus on meteorology was influenced by his awareness of how important weather forecasts were to the military and his understanding of the mathematics of the problem—he had done much work in fluid dynamics (Aspray, 1990, p. 130). According to Philip Thompson, von Neumann "regarded [the forecasting problem] as the most complex, interactive, and highly nonlinear problem that had ever been conceived of—one that would challenge the capabilities of the fastest computing devices for many years." (1983, p. 757).

What von Neumann most needed was a hitherto unsolved scientific problem that would yield to computational power and a problem of practical import to secure funding and impress the world. Meteorology served von Neumann well.

On the other side of the ledger, von Neumann certainly served meteorology well. He secured funding for the Meteorology Project.[15] He secured an institutional setting for the Project. He made arrangements in 1949 for the meteorologists to use the ENIAC at Aberdeen Proving Ground, and it was on the Institute computer that the first successful numerical forecasts were made. He aroused the interest of talented people in numerical forecasting, and he worked to involve them in the work of the Project, either as consultants or as full-time participants.[16] There cannot have been many people in the 1940s better able than John von Neumann to secure these financial, institutional, instrumental, and intellectual resources.

Early in the year 1946 von Neumann began studying the meteorological literature, and since he had earlier done much work in fluid dynamics, he soon, according to Thompson, "developed a good feeling for what the problem was."[17] Harry Wexler, a leading meteorologist of the Weather Bureau who frequently visited Princeton, wrote in September 1946 that "von Neumann is more than an adjunct to the computer" and that "He has considerable insight into physical problems and he has acquired much meteorological knowledge through extensive reading in the past few months."[18]

Though engaged in other work, particularly consulting, which often took him away from Princeton, von Neumann gave much of his time to the meteorology project. He sometimes set specific tasks for the meteorologists, and he frequently solved mathematical problems for them (Aspray, 1990, p. 142).[19] His skills in programming and numerical analysis were a continual help. And he acted as a channel between the computer designers and the meteorologists: he told the meteorologists what computer performance they could expect, and he told the computer designers what capabilities were most important to the meteorologists. Julian Bigelow, the first chief-engineer of the Institute computer, has written

> Johnny often emphasized that the big performance payoff would be in multiplication speed, and asked how fast we thought we could make this instruction function. If one made a thoughtless estimate, crises would arise at lightning speed, for the estimate would be used by him as a basis for a vast structure of deductions about what sorts of calculations would or would not be feasible in consequence of such speeds.(1980, p. 293)

So although it must be admitted that von Neumann's vital contribution to numerical meteorology was as impresario, he was also a director and a major performer.

The Start of the Meteorology Project

There were in 1945 a few besides Zworykin and von Neumann who thought the time might be ripe for numerical forecasting. One was a young meteorologist,

Robert D. Elliott, at the California Institute of Technology. Two years earlier Elliott, emulating Richardson, had devised an algorithm to predict pressure changes and had made a trial forecast using the algorithm. He submitted in 1943 some of his results to the Weather Bureau in a manuscript entitled "Calculated weather forecasts," but this seems not to have attracted much attention.[20]

Elliott's approach was to modify and greatly simplify Richardson's scheme. He had, as well, more modest expectations: "not high accuracy, but correct orders of magnitude." Elliott concluded[21]

> The results in every way verified the hypothesis tested by this experiment, namely, that a computing scheme could be set up which would yield a reasonable result. At the surface most of the predicted 12 hour pressure changes were of the correct sign and all compared reasonably well to the observed.

Elliott admitted that standard forecast procedures were more accurate, but thought that by the use of a finer grid and the inclusion of more terms in the equations one might get quite accurate forecasts. He did not, in this manuscript, address the issue of carrying out the computations fast enough for them to be of practical use, although it is likely that this concern partly explains his concluding statement: "The development of this method is of such magnitude that it is probably too difficult to be solved without help from outside the field of meteorology."

The outside help came in the spring of 1946, and it was Carl-Gustaf Rossby, a great entrepreneur of meteorological ventures, who got things moving. After visiting von Neumann in Princeton he wrote a long letter to Francis Reichelderfer, Chief of the Weather Bureau, remarking, "In view of Professor Von Neumann's outstanding talent, it would obviously be desirable for us to encourage his continued interest in meteorology."[22] He described the meeting with von Neumann and his own thoughts on the feasibility of numerical forecasting. This letter, which was read and commented on by several people at the Weather Bureau, recommended that a project be started and suggested personnel:

> It seems to me that an agreement should be reached between the Institute for Advanced Studies in Princeton, and an approved government agency in Washington—for instance, the Office of Research and Invention of the Navy Department to permit Professor Von Neumann to organize this preliminary group by offering suitable positions to Pekeris, Hunt, and Montgomery or Panofsky. I assume that RCA Laboratories would be willing to make Dr. Elsasser available. . . .

After talking to Harry Wexler, director of research at the Weather Bureau, and Daniel F. Rex of the Navy's Office of Research and Invention, Rossby wrote to von Neumann on 23 April encouraging him to submit a proposal for a meteorology project (Aspray, 1990, p. 132). On 1 May 1946 von Neumann met with Rossby and the French meteorologist Paul Queney, and a week later von Neumann sent a detailed proposal (seven typewritten pages, single-spaced) to Commander Rex.[23]

Von Neumann wrote that the Institute computer would not be completed for 2 or 3 years, but that the Meteorology Project should begin immediately since much work had to be done before such a computer could be of much value to meteorologists.[24] Von Neumann asserted that "efforts in meteorological theory were in the main limited by what was practical in actual computing," so that with the prospect of calculational methods 1000 to 100,000 times faster than what had been possible a "complete reassessment or revaluation of the theory" was "an absolute prerequisite." He wrote

> A careful analysis of the present status of meteorological theory . . . indicates that even if computing equipment of the type indicated above were immediately available, we would not be able to use it at once. . . . Indeed, the possibilities that are opened up by these devices are so radically new and unexpected that the theory is entirely unprepared for them.

Von Neumann proposed that this reassessment be carried out by 6 full-time meteorologists, assisted by 10 consulting meteorologists and two clerk-computers, with a total yearly budget, including salaries, rental of office space, and travel expenses, of $61,000.

The proposal was accepted with the official starting date of 1 July 1946. Von Neumann, with the assistance of Wexler, contacted a number of meteorologists and got several to agree to work with the Project,[25] but not much happened in the first months. Von Neumann called a meeting, which took place at Princeton on August 29th and 30th, that was attended by about 20 meteorologists, including Rossby, Queney, Wexler, and Jule Charney.[26] This meeting stimulated a number of meteorologists to begin thinking about numerical weather prediction,[27] and, according to Wexler, "a good deal was accomplished in determining the major problems in dynamic meteorology to be tackled by the computing group."[28]

Nevertheless, this meeting did not set the Project in motion as von Neumann had hoped. The major difficulty was staffing. There were not many meteorologists with the necessary training in physics, and of these not many could be induced to make even a part-time commitment to the Project. Making the problem worse was a very serious shortage of office space and housing in Princeton.[29]

In mid November 1946 when Philip Thompson, a young Air Force meteorologist, joined the Project, the only meteorologists resident at the Institute were Paul Queney, Chaim Pekeris, and Gilbert Hunt, and only Queney was working full time for the Project. Albert Cahn, who had been appointed director of the Project, withdrew for personal reasons shortly before Thompson's arrival. (Panofsky at New York University and Wexler of the Weather Bureau were nonresident consultants.) So severe was the staffing problem that at the end of October von Neumann told Wexler that "if conditions did not improve he would seriously consider giving up the project."[30] It may have been Thompson's presence that kept the Project going.

In the next year and a half the staffing situation did not improve. Hunt was working on a Ph.D. in topology and could give little time to the Project. Queney returned to France, and Pekeris left for Israel. So for quite a while Thompson was the only meteorologist associated with the Project who was working in Princeton, and he recalls that he "felt pretty isolated there being all alone (Thompson, 1986, p. 4).

The work of the Project in these first 2 years seems to have been, for the most part, a variety of separate studies. Queney worked mainly on the perturbations of air flow caused by the ground profile; he classified the waves thus engendered and investigated their stability properties. Pekeris investigated the stability of plane Poiseuille flow by numerical experimentation; von Neumann reported that the "numerical difficulties in this domain are very considerable."[31]

Panofsky and von Neumann collaborated to develop algorithmic methods of weather-map analysis, in particular, fitting "streamlines, isobars, and other characteristic curves to the observed wind and pressure maps."[32] Adapting their method to the calculating aids available, Pekeris and von Neumann devised an algorithm that involved a 10-parameter fit to the observational data. They then used a 10-equation, 10-variable electrical linear-equation solver that RCA had developed. This was an analog computer and its accuracy, three significant figures, was not sufficient.[33] Pekeris and von Neumann then achieved some success with digital methods, securing for this purpose the assistance of the New York-based Mathematical Tables Project of the National Bureau of Standards.[34] Von Neumann studied numerical methods of integration of the hydrodynamic equations describing the atmosphere, and Thompson made a study of "Permanent waves in homogeneous, plane fluids."

Finally, an attempt was made to produce forecasts, using real data, by numerical integration. But even with the computational assistance of the Weather Bureau this attempt seems to have been abandoned. In May 1948 von Neumann reported[35]

> The routine of numerical calculations will require on the order of 500 calculator-hours per time stage; since the Weather Bureau has been able to furnish only 70 man-hours per week, it is imperative that the remainder of the computations be carried out by an agency specially equipped to handle them. . . . A proposal to procure the computational facilities of the Bureau of Standards will be forwarded separately.

The attempt was not mentioned in subsequent progress reports.

So in 1947 and early 1948 the Meteorology Project, although producing a number of valuable individual studies, was neither working as a team nor making much progress toward the goal of numerical weather prediction. This picture was changed by the arrival in mid 1948 of Jule Charney.

The Arrival of Jule Charney

Charney received his bachelor's degree from the University of California at Los Angeles in 1938. He continued at UCLA as a graduate student in mathematics, earning in 1940 an M.A. in mathematics, but in 1942 he changed fields, partly as a result of the war, to meteorology.[36] In 1946 he completed his Ph.D. and was awarded a 1-year National Research Council Fellowship, which he planned to use at the University of Oslo. On his way to Norway he stopped at the University of Chicago, intending to stay only a few days. The intellectual atmosphere there was so stimulating that Charney postponed the start of his fellowship and remained there 9 months. Charney later said of this period at Chicago, "it was . . . the main formative experience of my whole professional life" (1987, p. 62). This effect was principally due to the presence of Rossby, with whom Charney had "*endless* conversations." In the spring of 1947 Charney traveled on to Oslo, where he worked mainly with Arnt Eliassen. Later that year he received an invitation from von Neumann to join the Meteorology Project and the following July arrived in Princeton.

Charney was well informed about the Institute's project from the outset. He had attended the meeting in Princeton in August 1946 and the conference in Chicago that December (which was attended also by Queney, Thompson, Wexler, and Panofsky, all working with the Project). On his way to Norway in March 1947 Charney stopped for several days in Princeton, where he talked with Thompson and von Neumann. And in November of that year he said in a letter to Thompson that he had "been brooding about the probl[em] of numerical computation ever since coming to Norway" (in Aspray, 1990, p. 136).

In Chicago and Oslo, Charney studied how the basic equations describing the atmosphere might be modified so that the only solutions to the equations would be the meteorologically important ones and so that the equations could be solved numerically. Charney thought the first desideratum could be achieved by imposing a filter, as it were, on the motions described by the equations. The basic equations, such as the set of equations Richardson used (Figure 4 of Chapter 6), can be said to describe too much since they admit solutions that correspond to types of motion that are not meteorologically significant, especially higher-frequency wave motions such as sound waves and gravity waves.[37] The problem is that the complete solution of the equations is greatly complicated by the existence of solutions that are meteorologically insignificant. Charney worked to alter the equations so that these solutions are eliminated while retaining the meteorologically significant ones.[38] He found that the geostrophic approximation—motion of the atmosphere is said to be geostrophic when the pressure-gradient force is exactly balanced by the Coriolis force—"could be used to reduce the equations of motion to a single dynamically consistent equation in which pressure appears as the sole dependent variable" (Charney, 1951, p. 470).[39]

The second desideratum was that the equations be numerically solvable. Recall that Richardson replaced the differential equations describing the atmosphere by difference equations. Instead of a derivative (which may be thought of as an infinitesimal change divided by an infinitesimal change), he computed the quotient of two finite differences (where the differences are between values at adjacent grid points or at successive times). He assumed that as the three-dimensional grid became finer and as the time increments became smaller, the solutions of the difference equations would approach the solution of the differential equation. In a 1928 paper, which became one of the most cited papers of the 20th century, Richard Courant, Karl Friedrichs, and Hans Lewy showed that this does not always happen.

They showed in particular that for an important class of partial differential equations a certain inequality had to be satisfied in order for the solutions of the difference equations to converge to the desired solution.[40] When this "Courant condition" is not satisfied the procedure is said to be computationally unstable. It turns out that Richardson's algorithm does not satisfy the Courant condition, so that its repeated application would not give useful results.[41] Charney was aware of this and found that by filtering out most atmospheric motions he made it much easier to satisfy the Courant condition. In Thompson's words,[42]

> Charney's 1947 formulation, later known as the quasi-geostrophic model [motion is said to be quasi-geostrophic when there is an *approximate* balance between the pressure-gradient force and the Coriolis force], simultaneously skirted two major difficulties: first, it imposed much less stringent conditions for computational stability, and second, it did not demand that the horizontal divergence or accelerations be computed as small differences between large and compensating terms, each of which is subject to sizable percentage errors. These features alone evaded the two fundamental difficulties inherent in the practical application of Richardson's method. (1983, p. 760)

In late 1947 von Neumann wrote to Charney inviting him to become director of the Meteorology Project. On 2 January 1948 Charney accepted and requested that Arnt Eliassen be invited to join the Project for 1 year.[43] Charney wrote that Eliassen would be especially valuable to the Project because of his knowledge of synoptic meteorology as well as of dynamic meteorology. Later in the letter Charney again urged that the Project increase its contact with empirical meteorology, both by making use of observational data and by having Eliassen spend time at the University of Chicago and at the Weather Bureau.

After Charney's arrival in July 1948 the work of the Project became more unified and directed. This is evident in the appearance of the progress reports (which were prepared about twice a year): those written before July 1948 suggest a group of independent researchers, those written after suggest a coherent research project. Jerome Namias, a meteorologist with the Weather Bureau in Washington who made frequent trips to Princeton, wrote, "Early on, it became clear that since Jule Charney had appeared in Princeton, the project was headed in the right direction and practically important results might take place within a few years," and "The Princeton group was a wonderfully organized team—a pleasure to see in action" (Namias, 1986, p. 21).[44]

Charney's Program

In 1952 Charney wrote, "The philosophy guiding the approach to this problem has been to construct a hierarchy of atmospheric models of increasing complexity, the features of each successive model being determined by an analysis of the shortcomings of the previous model" (in Smagorinsky, 1983, p. 13). As this quotation suggests, Charney insisted that the approach to useful numerical forecasting be through implementable algorithms and in small steps. In other words, he insisted that the development occur in the realm of computational feasibility and that changes be made one at a time. One starts with a model—necessarily drastically simplified—whose behavior can be computed. One then judges that model according to its predictions and alters the model, usually by introducing a physical factor whose earlier exclusion is thought to have been a cause of the discrepancies between predicted and observed motions.[45] Its predictions are then evaluated, and so on.

One might say that Charney felt "theory pull" exceptionally strongly, since he invariably took pains to connect theoretical models to observational evidence. Norman Phillips has written, "I believe this aspect of Jule—the importance he attached to comparing theory with reality, and his awareness of the care that this often requires—is something we should be even more grateful for than we are for his well-known theoretical ability" (1982, p. 494).[46]

For a number of reasons Charney's program was a success. A vital factor was the high quality of his co-workers. Also vital was the fact that the realm of computational feasibility grew vastly in this period: from hand computation with desk calculators to the ENIAC to the IAS Computer to the IBM 701. And what got things moving was Charney's excellent starting point, the quasi-geostrophic model, according to which important motions of the atmosphere are described by a single equation.

The computational difficulties were nevertheless formidable. Although atmospheric pressure was the only dependent variable in the quasi-geostrophic equation, there were the four independent variables of position and time, and this made the problem too complex for the available means of computation. Charney and Eliassen eliminated one variable by "tak[ing] advantage of the fact that [the] large-scale atmospheric motions exhibit a kind of degeneracy with respect to the vertical coordinate";[47] they ignored the vertical variation in motion and assumed that the vertically averaged horizontal motion of the atmosphere was the same as the horizontal motion at a particular height, called "the equivalent barotropic level."[48] Charney (1949) showed how to reduce the model further to a single spatial dimension, namely a latitude circle. Charney and Eliassen (1949) then used a linearized version of this equation to make the first numerical predictions, which they were able to do by hand computation.[49]

Figure 1 below shows how two of their 24-h forecasts compared with the observed values. The variable Z is the height of the 500-mb surface; its predicted value along 45° north latitude from 180° west longitude to 180° east is shown at the top. The predicted changes (solid line) are compared to the observed changes (dotted line) at the bottom.

Charney and Eliassen devised ways to include the effects of friction and topography in these predictions. They found, however, that for 24-h forecasts these factors had little effect. This finding was later used as evidence that friction and topography could be ignored in the first attempts at short-range prediction.[50]

The authors believed that the method was sufficiently accurate to justify its use in day-to-day forecasting (Charney and Eliassen, 1949, p. 38). Rossby thought the method was "extraordinarily promising" and wrote on 8 May 1949, "Namias, who is here, got extremely excited [about] the whole thing [that is, the linear one-

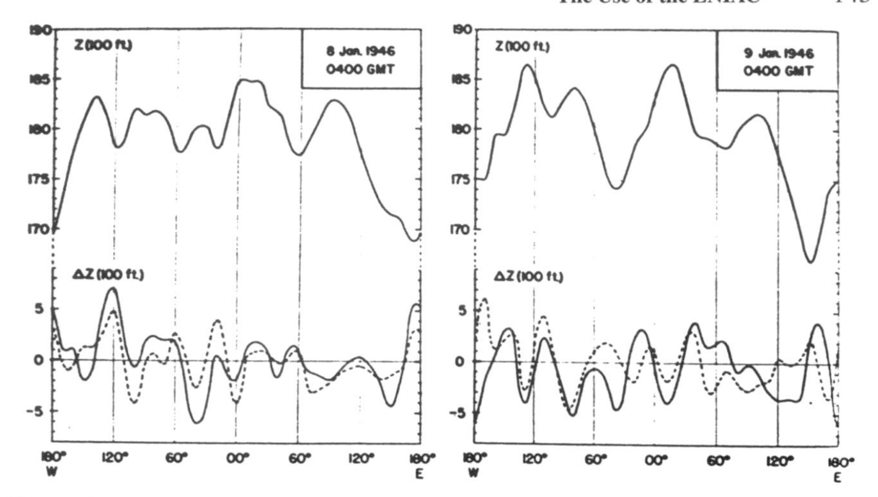

Figure 1 Representations of two of the trial 24-h forecasts reported in Charney and Eliassen's article "A numerical method for predicting the perturbations of the middle latitude Westerlies" (Charney and Eliassen, 1949).

dimensional model of Charney and Eliassen] and wrote to Harry Wexler to have him interview Charney with the idea of introducing the method in the Weather Bureau. . . ." (in Platzman, 1979, p. 308). The Meteorology Project made further tests of the procedure, and independent tests were conducted at Stockholm under Rossby's supervision[51] and by J. Sawyer and F. Bushby of the British Meteorological Office. Sawyer and Bushby (1951) were less enthusiastic than Rossby and Namias, reporting that

> The one-dimensional method of computing the change in the 500 mb height profile given by **Charney** (1949) has been applied to various latitudes on a total of about 60 charts. Significant success was achieved. . . . However the success was definitely less than that achieved by conventional forecasting methods on the same charts. (p. 201)

It seems that the method did not become a part of standard forecasting procedure. It did, though, make many people think that successful numerical forecasting was just around the corner: if so simple an algorithm can do fairly well, then the much more sophisticated algorithms, soon to be carried out by electronic computers, will certainly do excellently.[52]

The Use of the ENIAC

The first numerical forecasts were made with a one-dimensional model in which the equation of motion was linearized. Charney and Eliassen decided next to investigate a nonlinear two-dimensional model. In June 1949 they believed that the

Institute computer would soon be available for doing the computations, and they proposed to carry out a few time steps by hand and desk calculator as preparation. By November it was clear that the Institute computer would not be completed in the next year and that the hand computations were too difficult.[53]

Plans were therefore made to use the ENIAC, which had earlier been moved from Philadelphia to the Army's Ballistic Research Laboratories in Aberdeen, Maryland. Von Neumann and Reichelderfer arranged for the Meteorology Project to have the use of the computer in December, but preparations took longer than planned and the first ENIAC "expedition" actually took place in March and April 1950.[54]

By Charney's account the preliminary work included finding (1) a method of solution suitable to the ENIAC, (2) appropriate boundary conditions, (3) computational stability criteria, and (4) limits on the area into which influences from the boundary of the forecast region propagate.[55] These problems were "attacked and solved, principally by von Neumann." By March the work of programming and coding also was complete. As Platzman remembers it, "The proceedings in Aberdeen began at 12 p.m. Sunday, March 5, 1950 and continued 24 hours a day for 33 days and nights, with only brief interruptions" (Platzman, 1979, p. 307). Besides von Neumann (who did not make the trip to Aberdeen) and Charney, the main meteorologist participants in this work were Ragnar Fjörtoft, John Freeman, George Platzman, and Joseph Smagorinsky.

ENIAC had 20 registers, each capable of storing a 10-digit number, but these registers acted both as internal memory and arithmetical unit so that in general fewer than 10 registers were available as read–write memory to the programmer. There was read-only memory in three so-called "function tables," each holding 208 6-digit numbers or 104 12-digit numbers (Williams, 1985, p. 277). Input and output were by means of punched cards, and because of the extremely limited direct-access memory it was necessary to make continual use of punched cards as external memory.

Figure 2, which is Platzman's diagram of the 16 operations for each time step, gives some impression of the awkwardness of using the ENIAC for solving the prediction equation. Notice the large number of punched card operations; in the course of the ENIAC expedition 100,000 punched cards were produced (Charney *et al.,* 1950, p. 245). The magnitude of the computation was impressive: von Neumann said, "It would have taken five hand-computer-years to have duplicated the Eniac computations."[56]

However, even with a 1-week extension of the 4-week period of access to the computer, the Project was able to carry out only four 24-h forecasts and two 12-h forecasts. Malfunction of the ENIAC and programming difficulties were the main causes of delay.[57]

The forecasts were surprisingly good. Charney wrote, "The results showed . . . that with certain well marked exceptions, the large scale features of the 500 mb flow can be forecast barotropically. . . . All in all I think we have enough evidence

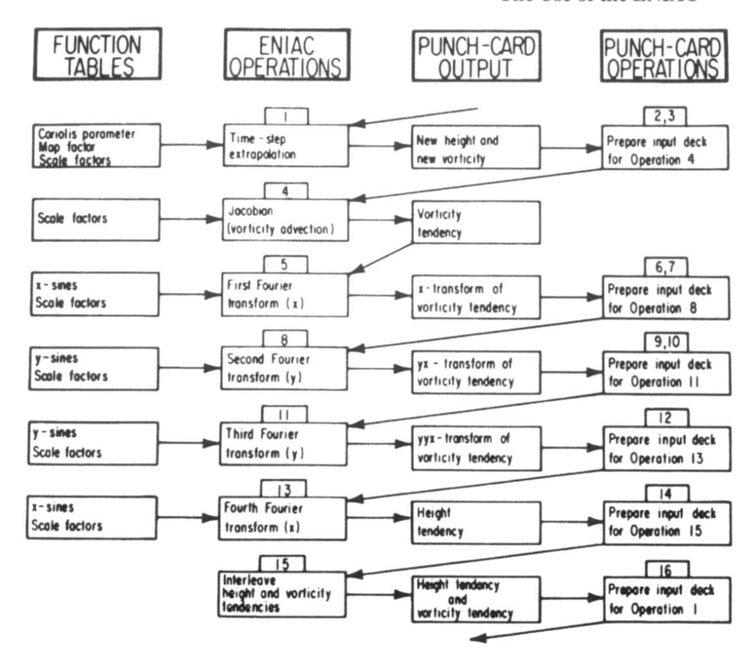

Figure 2 A diagram of the 16 operations in each time step of the forecasts made in the first ENIAC expedition (© American Meteorological Society. From Platzman, 1979, p. 309).

now to bear out most of Rossby's prophecy [that "we are standing at the threshold of a new era in applied meteorology"]" (in Platzman, 1979, p. 312).

The results of the four 24-h forecasts were presented and analyzed by Charney, Fjörtoft, and von Neumann in the article "Numerical integration of the barotropic vorticity equation," which appeared in *Tellus* in November 1950. They reported, "The computation time for a 24-h forecast was about 24 hours, that is, we were just able to keep pace with the weather." They estimated that with the Institute computer "the total computation time with a grid of twice the Eniac-grid's density, will be about 1/2 hour." Hence, for the first time meteorologists had good reason to believe that they would soon overcome both of the fundamental difficulties described at the beginning of this chapter—to devise an algorithm that does as well as human forecasters and to carry out the algorithm fast enough for the forecasts to be useful.

There was a second ENIAC expedition. Charney, Platzman, Phillips, Bolin, and Smagorinsky were involved. The purpose was to model barotropic instability.[58] There was no clear-cut outcome of the trials made, and no report on this expedition was ever published. Nevertheless, it is possible, according to Platzman, that these computations are historically important for being the first time that a certain type of nonlinear computational instability was observed.[59]

One of the simplifying assumptions made in devising the algorithm implemented on the ENIAC was barotropy: that the surfaces of constant pressure (in the atmosphere) do not intersect the surfaces of constant density. When such surfaces intersect, which is the usual case, the atmosphere is said to be baroclinic. In a barotropic atmosphere there is no conversion of potential energy to kinetic energy. Since the generation and intensification of storms were believed to depend on such an energy conversion, it was expected that the barotropic model could not predict such motions.

Although this expectation was on the whole borne out by the ENIAC forecasts, there were some atmospheric developments, thought to be essentially baroclinic, that were correctly predicted. Charney *et al.* saw the forecasts as evidence that "the effects of baroclinicity do not manifest themselves in a steady, widespread conversion of potential into kinetic energy, but rather in sporadic and violent local overturnings accompanying what, for want of a better term, may be called baroclinic instability, and second, that when these effects are not predominant, the motion is quasi-barotropic" (Charney *et al.,* 1950, p. 238). They wrote also, "One has the suspicion that certain processes which have heretofore been classed as baroclinic will be found to have a barotropic explanation."

Nevertheless, the algorithm used for the ENIAC forecast was not seen as an end in itself but as a step in Charney's program: "Just as the analysis of the linearized barotropic equations served as a pilot study for the integration of the non-linear barotropic equations, so will these integrations supply the necessary background for the treatment of the three-dimensional equations" (Charney *et al.,* 1950, 238). Indeed, Charney later said that he was then imbued with the idea "that the prediction of cyclogenesis . . . was the major problem of meteorology" (1987, p. 105).

The Use of the IAS Computer

Charney, Fjörtoft, and von Neumann in the 1950 paper describing the ENIAC forecasts wrote, "The most typical conditions for the breakdown of the barotropic model will probably occur when potential energy is converted into kinetic energy," (Charney, 1987, p. 250), and the final section of the paper is a consideration of a simplified baroclinic model. As Thompson reports it, "there was a general rush to develop baroclinic models":

> In a relatively brief span, 1951–53, no less than six simple baroclinic models were proposed, two of which were tested in a few real cases: all were variants of the general quasi-geostrophic model developed by Charney in his paper of 1949. (1983, p. 761)

Since vertical motion of the atmosphere is often involved in the conversion of potential energy to kinetic energy—as when warm air overrides a wedge-shaped body of cold air—many of these models represented the atmosphere as a number of layers so that vertical motion could to some extent be modeled. In 1951 Norman

Phillips published the article "A simple three dimensional model for the study of large-scale extratropical flow patterns" in which he presented a two-layer model, which came to be called a $2\frac{1}{2}$-dimensional model.[60] Phillips, who was working on his Ph.D. at the University of Chicago, became a consultant to the Institute project in May 1951 and a full-time member several months later.

It was not until the spring of 1952 that the Institute computer became available to the Meteorology Project.[61] The first tests run on the computer were barotropic forecasts of the famous Thanksgiving Day storm of 1950. That storm became, for meteorologists, a favorite object of study because it developed extremely rapidly and was unusually severe.[62] It was regarded as a prime example of baroclinic development, and for that reason Phillips had, in his 1951 paper, tested his baroclinic model against it. The Meteorology Project made the barotropic forecasts of it in order to compare these forecasts with the baroclinic forecasts they planned soon to make.

Charney and Phillips devised a two-layer model that was then programmed for the Institute computer. There were considerable problems with computer malfunction, and input and output (using teletype equipment) consumed more time than did the computations, but by 25 August 1952 the first 12-h forecast was completed.[63] Correlation coefficients for the predicted and observed 12- and 24-h changes were computed for each of the models. The results are shown in Table 1 below.

Table I

Correlation Coefficients for Predicted and Observed Changes of Pressure for Two Models, Barotropic and Baroclinic, in the Period Including the Thanksgiving Day Storm of 1950[a]

	Barotropic forecast for 500 mb		2 Layer-model forecast for 700 mb	
Initial date	12 hour forecast	24 hour forecast	12 hour forecast	24 hour forecast
Nov. 23/50, 0300Z	.89	.72	.89	.90
23/50, 1500Z	.81	.77	.86	.87
24/50, 0300Z	.88	.75	.86	.73
24/50, 1500Z	.74	.51	.77	.62
25/50, 0300Z	.61	.61	.58	.56
25/50, 1500Z	.81	.59	.73	.87

[a] Progress Report Calendar Year 1952.

Problem	Programmers	Status	Remarks
1. Gen'l quasi-geostrophic model; vert. coord. p; 5 Levels; vortex tube rotation in vertical plane ignored. Horizontal ground.	Gilchrist Lewis Cooley	Conversion and reconversion codes completed and checked. About 2 weeks remaining on main code.	To be run on I. B. M. 701.
2. Same as 1, but vortex tube turning not ignored.	Gilchrist Lewis Cooley	Mathematically formulated. Coding not begun.	To be run on I. B. M. 701. Liebmann process applied to elliptic eq. with mixed terms.
3. Gen'l quasi-geostrophic model; vert. coord.	Phillips Lewis Gilchrist	Code completed and partially checked out. Needs some revision.	
4. Same as 1, but vertical advection of pot. vort. ignored.	Blackburn Schuman	Flow diagram completed. Coding begun.	Integration much simpler than for prob. 1. Possible model for operational prediction.
5. Same as 4, but vortex tube turning not ignored.	Blackburn Schuman	Mathematically formulated. Coding not begun.	Liebmann process applied to elliptic eq. with mixed terms.
6. 2-dimensional objective analysis.	Stickles Cooley Gilchrist	Code completed but not checked out. Needs revision.	Stickles will probably not be able to run. Gilchrist or Cooley should take over.
7. Computation of vertical velocities from 3-level predictions with semi-linear model.	Blackburn Schuman	Code completed but not checked out.	
8. Finite amplitude wave and vortex development, jetogenesis, in barotropic model.	Gilchrist Lewis	Models more or less decided upon. Math. formulated. Coding not begun.	Requires alteration of bound. conds. in existing barotropic code. In part a continuation of 1951 Eniac work.
9. Finite amplitude wave and vortex development, frontonogenesis, jetogenesis, forced meridional circ., in 2-layer model.	Gilchrist Lewis	Same as 8.	Requires alteration of bound. conds. and general revision of existing 2-layer code.
10. Integration of primitive eqs. for 1-layer atm. with free upper boundary	Bolin	Code completed and partially checked.	Will try various signal speeds $\sqrt{gN}$ to test inertial-gravitational interaction.
11. Integration. of primitive eqs. for 2-layer atm. with rigid upper boundary.	Veronis	Math. formulated. Coding not begun.	Object -- to compare results with quasi-geostrophic model.
12. Computation of spectra and co-spectra of turbulent velocities at low levels in atmosphere.	Nussbaum	Code completed and checked out.	Computation for Panofsky. Believe computations could be greatly speeded up by code revision.
13. Integration of barotropic equations by Fourier representation.	Kady	Code partially written.	
14. Study of Comette flow by calculation of motion of point vortices.	Fjørtoft	Initial vortex configuration decided. Coding not begun.	
15. Tests of truncation error in barotropic model.	Gilchrist	Requires slight rescaling and change in existing code.	
16. Test of convergence properties of Liebmann process in 3-level semi-linear model.	Gilchrist	Requires negligible change in existing code.	
17. Integration of 2-layer model with externally given dh/dt and frictional dissipation for long periods.	(?)	Mathematically more or less formulated except for treatment of trunc. error.	

Figure 3 "Computing program for meteorology" at the Institute for Advanced Study in November 1953 (von Neumann Papers). This chart was probably drawn up by Charney.

Notice that the 24-h barotropic forecasts were markedly less accurate than the 12-h barotropic forecasts. The baroclinic forecasts, on the other hand, maintained a high correlation. The progress report for calendar year 1952 concludes

> Even with the present crude models, cursory comparison with subjective prognoses made by experienced forecasters indicates at least comparable accuracy. Moreover, whereas subjective methods have not shown significant improvement in the past 20 years, the present approach may be refined in a logical manner.

These baroclinic forecasts were widely seen as evidence that numerical forecasting could outperform subjective forecasting.

Early the following year Charney and his associates used a three-layer model to make the same forecasts, and the correlation coefficients were significantly higher. Further tests were done on other three-layer models and on two five-layer models.

In 1952 and 1953 the amount of activity of the Meteorology Project seems to have increased markedly. There was certainly a sharp increase in the number of full-time participants, consultants, and visitors. The progress report for 1 July 1953 to 31 March 1954 lists 16 meteorologists who were at the Institute for all or part of that period. Figure 3 gives some indication of the amount and range of activity late in 1953.

There seems to have been a falling-off in activity in 1955. In 1954 von Neumann became a member of the Atomic Energy Commission and was therefore rarely in Princeton, and in 1955 failing health—he died of cancer in early 1957—began restricting his activities. There was no other permanent member of the Institute interested in the Meteorology Project. Moreover, many of the members of the Institute felt strongly that the Institute was not the place for experimental or applied research, and Charney later said, "[This feeling] was probably the main reason I left the Institute. . . ." (1987, p. 96). In 1956 Charney and Phillips accepted positions at MIT, and the Institute terminated the Project. By this time, however, the Project had done much to establish a new style of meteorology, which is the subject of the following chapter.

Chapter 11 The Acceptance of Numerical Meteorology

The Beginnings of a New Style of Meteorology

A new style of meteorology evolved in the 1950s. It was much the style practiced by Richardson some 35 years earlier, a style that results from making a forecasting algorithm one's ultimate objective. This style is characterized by interest in only certain types of data (quantitative, and with accuracy and geographic distribution determined by the needs of the algorithm), interest in only certain types of theories (quantitative, and "meshing" with other quantitative theories, and dealing with only those phenomena the algorithm indicates are important), and interest in numerical analysis and computing aids. In the 1950s it was the availability of electronic computers that turned many meteorologists in the direction taken by Richardson much earlier.[1]

This redirection began even before stored-program computers were available, the mere prospect of computers exerting some force. In recounting the early history of numerical meteorology, Charney wrote of "the great psychological stimulus that the very possibility of high-speed computation brought to meteorology," and in 1949 Rossby reported, "In the last year or two a good deal of attention has been given to the problem of formulating the basic equations of atmospheric hydrodynamics in such a manner that they may be integrated numerically with the aid of high-speed computing equipment" (Charney, 1972, p. 117; Rossby, 1949, p. 54).

There was renewed interest in Richardson's work. Beginning in the mid 1940s his name appears more and more frequently in the meteorological literature, and in 1950 Charney wrote, ". . . Richardson's work has ceased to be merely of academic interest and has become of the greatest importance. . . ." (p. 235).[2] New issues came to the fore, an example being that there was new interest in the speed of propagation of atmospheric effects, since, as Rossby points out, "this speed obviously determines the size of the area over which initial conditions must be known if numerical predictions are to be made for a prescribed locality" (1949, p. 54).

The data requirements of forecasting algorithms began to have some influence on observational efforts. A 1952 article written by staff members of the Institute of Meteorology in Stockholm directly addresses this issue: "Data requirements for this kind of computations are discussed on the basis of the difference in the success of the method over regions with poor and with good data," and the authors recommend an increase in radio wind observations (pp. 21, 30).

Another important effect of computers was the facilitation of numerical experimentation, which in many cases was not connected with any forecasting algorithm, an example being Pekeris's work described above. And one may say that von Neumann's vision of heuristic computation was realized in the 1960s when numerical experimentation came to be a standard methodology for meteorologists. George Cressman wrote in 1972, "The development of the electronic computer changed everything. By permitting experimentation, it was responsible for new extensions of theory, which advanced in parallel with their practice" (p. 181). This effect of the availability of computers is discussed more fully in Chapter 12.

The work of Charney and others at the Institute had made clear the value of electronic computers in meteorology. It made clear too how meteorologists were constrained by computer hardware: by limited reliability, by limited speed, by limited memory, and by the restriction to arithmetical operations.[3] According to Platzman, the ENIAC "had a mean error-free path of only a few hours and often took many hours to repair"(1979, p. 310),[4] and the Institute computer also proved troublesome. Speed was a continual concern. During the first ENIAC expedition Platzman wrote, "we find that the ENIAC is a little slower than was anticipated so that some recoding has to be done," (1979, p. 310), and the published report of the 24-hour forecasts, which took 24 hours to compute, considers ways to speed the process (Charney *et al.*, 1950, p. 245).[5] The memory capacity of the computer was sometimes decisive in the selection of an algorithm, as Charney makes clear in the following passage:

> There are a number of methods, both analytic and numerical, by which (18) can be solved. The best for hand calculation is probably the relaxation method of Southwell [14] as it is rapid and can be used with any type of boundary. However this method is not always suitable for automatic machine computation as it makes large memory demands. It has been found best for a machine with a small internal and large external memory to use the analytic solution expressed in terms of a Green's function . . . . (1951, p. 472)

The restriction of a computer algorithm to logical and arithmetical operations was a constraint meteorologists were not accustomed to. In late 1947 the meteorologist Victor Starr wrote to von Neumann to ask whether the computer being developed could be used to solve some equations he had derived. Von Neumann replied,[6]

> . . . it will only solve problems which are resolved into arithmetical operations and explicit procedures. Transcendental and implicit problems must, therefore, be transformed in this manner. This means usually that one has to resort to methods of stepwise integration, successive approximation by iterations, successively self-correcting, trial and error procedures—and combinations of these.

Even when the reduction of an algorithm to logical and arithmetical steps was theoretically straightforward, it could be a frustrating and time-consuming task. In one progress report Charney wrote, "The majority of the Group's time in the third quarter of 1951 was spent in programming."[7]

Besides all these hardware constraints, the meteorologists were constrained also by the state of numerical analysis, which was at the time a very young branch of mathematics. For example, Charney wrote that a certain model was "adopted because of its simplicity and adaptability to numerical analysis" (Charney *et al.*, 1950, p. 252). What was happening, of course, was that meteorology was becoming more closely tied to engineering and to other branches of knowledge.

The Spread of Numerical Meteorology

By 1952 there were four sizeable research groups working on numerical weather prediction: the Meteorology Project at the Institute, the Atmospheric Analysis Laboratory of the Air Force Cambridge Research Laboratory, the Napier Shaw Laboratory of the British Meteorological Office, and the International Meteorological Institute of the University of Stockholm. For each of these groups the main task was the devising of forecasting algorithms.

There was a good deal of contact between these groups. Ties were particularly close between the Institute project and the work being done at Cambridge, Massachusetts, and at Stockholm. Thompson (1986, p. 9), who had been at the Institute before going to Cambridge to set up the Atmospheric Analysis Laboratory, maintained contact with the Institute project by making a trip to Princeton every 2 or 3 weeks. There was good communication between the Princeton and Stockholm groups; for example, in early 1951, both Rossby and Bert Bolin spent several months in Princeton. There was some direct contact between the Institute project and that of the British Meteorological Office; for example, in 1951 Charney visited the British group.[8]

Developments in Russia in this period, however, were apparently quite independent, and it is therefore remarkable how exactly these developments paralleled the work mentioned above. In the early 1950s the Soviet meteorologists used computers, STRELA and BESM, that were roughly equivalent to the computers, the IBM 701 and the IBM 704, used by American meteorologists (Phillips *et al.*, 1960, p. 607). The similarity of the Soviet and the non-Soviet work provides additional evidence for the effect of computers on research.

In 1960 in the *Bulletin of the American Meteorological Society* Norman Phillips, W. Blumen, and O. Cote reviewed the development of numerical weather prediction in the Soviet Union. Until this article appeared few American or English meteorologists knew much about the Soviet work (Smagorinsky, 1983, p. 5). The authors conclude that "Non-Soviet work has in general treated the same questions with similar results" and that "Soviet and non-Soviet meteorologists have in general both made the same type of numerical forecasts" (Phillips *et al.*, 1960,

Table I

Some Passages from Phillips *et al.* (1960) That Indicate the Similarity between Soviet and Non-Soviet Work in Numerical Meteorology

Quotation	Element of Similarity
The same technique [as Kibel used in 1940] was used by Philipps (1939) and later by Eliassen (1948).	Analytic technique
. . . Kibel introduced the important parameter ϵ; $\epsilon = V/fs \sim 0.1$. . . . the same non-dimensional ratio is also very important in the scale theory of Charney (1947).	Defined quantity
This paper [by Obukhov in 1949] . . . is the Russian counterpart of the well-known papers by Rossby (1938) and Cahn (1945).	Article
The results obtained by Obukhov in this anlysis of the linearized problem are the same as those obtained by Rossby and by Cahn.	Results of an analysis
It [= Equation 27] *is equivalent to the so-called "balance equation" proposed by Charney (1955) and by Bolin (1955). . . .*	Equation
Graphical methods of solving (31)-(36) were developed around 1951 by Buleyev and Yudin. . . . The techniques used are very similar to the scheme devised by Fjørtoft (1952).	Graphical method
As was the case with the machine computations in the U.S. (Charney *et al.*, 1950) these first Soviet calculations were made with a barotropic model.	Algorithm

pp. 603, 607). Table 1 reproduces a few passages from the article by Phillips, Blumen, and Cote to suggest how exact a similarity they saw.

The new style of meteorology required skills different from those meteorologists traditionally possessed. One could hardly devise implementable algorithms that would give useful information about the atmosphere without having some of the skills of a numerical analyst, a mathematical physicist, and a synoptic forecaster. As von Neumann expressed it: ". . . the meteorological and mathematical aspects probably cannot be separated and personnel familiar with both aspects are needed."[9] Charney stressed that the theoretical work, including physical theory and numerical analysis, take continual account of the empirical knowledge of the atmosphere.[10]

The sudden prominence of numerical meteorology soon affected the education meteorologists received as meteorological programs at universities came to include training in numerical forecasting. In 1953 both Thompson at the Massachusetts Institute of Technology and Platzman at the University of Chicago gave courses in numerical weather prediction (Thompson, 1986, p. 37; Smagorinsky, 1983, p. 22) and other universities soon followed suit.

In the 1950s the range of numerical meteorology increased greatly. At a conference in 1955 von Neumann described research in numerical meteorology as comprising three activities: short-term forecasting (over periods from 1 day to 14 days), long-term forecasting, and the modeling of the general circulation of the atmosphere.[11] He said

> In considering the prediction problem, it is convenient to divide the motions of the atmosphere into three different categories depending on the time scale of the prediction. In the first category, we have motions which are, in the main, determined by the initial condition—in which we may extrapolate the initial tendencies over a short period of time. In the second category, we have the opposite extreme, namely, motions which are practically independent of the initial conditions. In attempting to forecast such motions, we concern ourselves with traits of the circulation which, on the average, will always be present. . . . Now, between the two extreme cases, there is another category of flows. In this category, we are sufficiently far from the initial state so that the details of the initial conditions do not express themselves very clearly in what has developed. . . . Nevertheless, certain traits of the initial conditions bear a considerable influence on the form which the circulation takes. The problem of forecasting over periods from 30 to 180 days . . . concerns flows that fall into this category. (Pfeffer, 1960)

The work of the Institute's Meteorology Project had been directed mainly toward short-term prediction. Von Neumann thought that, in 1955, the general circulation could be profitably studied by calculation. Such work had, in fact, begun in 1952, and in 1955 Norman Phillips designed the first numerical model able to reproduce the main motions of the earth's atmosphere. Modeling of the general circulation soon became a major area of study; this is described in Chapter 12. And the area von Neumann saw as most difficult—the long-term forecasting—has indeed proved so.

Steps toward Operational Forecasting by Means of Computer

After visiting the Aberdeen Proving Grounds on 4 April 1950 to see the trial forecasts made on the ENIAC, Weather Bureau Chief Reichelderfer wrote, "What we saw convinced us that a new era is opening in weather forecasting—an era which will be dependent on the use of high-speed automatic computers." [12] It took some time before numerical forecasting was part of the Weather Bureau's daily operations: not until July 1954 was there a numerical weather-prediction group, and this group, the Joint Numerical Weather Prediction (JNWP) Unit, did not begin its daily forecasts until May 1955. The JNWP Unit grew directly out of the Institute project.

One of the most important events was a meeting, arranged by von Neumann, at Princeton on 5 August 1952 to determine whether the time had come to prepare for operational forecasting.[13] Attending the meeting were—besides those already at the Institute—representatives of the Weather Bureau, the Air Weather Service, the Geophysical Research Directorate of the Air Force, the Aerology Branch of the Navy, the Office of Naval Research, the University of Chicago, and the Air Research and Development Command.

Von Neumann opened the meeting by describing the work that had been done at the Institute. He obviously thought that this work justified starting a program to establish daily numerical forecasting, but he foresaw two major problems: lack of trained people and unreliability of computers. What was needed, he thought, were mathematician–meteorologists.[14] Charney said that the Institute project could provide, as he put it, "indoctrination in our methods," and he invited the Air Force, Navy, and Weather Bureau to send representatives. The invitation was accepted, and it was agreed that the representatives arrive in Princeton on 1 December 1952.[15] Platzman said that the universities could offer training in numerical meteorology and that he hoped the University of Chicago would soon do so.[16]

The other problem, computer unreliability, had repeatedly slowed the work of the meteorologists at Princeton. At the meeting von Neumann said

> A technological problem exists because the preparation of forecasts on a routine basis will require that some computing machine be in perfect condition and operable at a given time of day every day of the year. Of the several very fast machines now in existence I do not believe any are capable of running at maximum efficiency for more than 50% of the time.

This problem he thought would be "partially solved in the next several years," whereas the solution of another problem—the memory limitations of computers—would take longer.

In early 1953 the solution of the training problem seemed to be in sight, and with the availability of a new computer, the IBM 701, the pace of the movement toward operational forecasting quickened. The 701, which was formally unveiled in the spring of 1953, was comparable in computing power to the Institute's machine but somewhat faster. On 4 June representatives of the Weather Bureau and the Air Weather Service met with representatives of IBM, and on 23 June the Ad Hoc Committee on Numerical Weather Prediction of the Joint Meteorological Committee made plans for a Joint Numerical Weather Prediction Unit to begin operations 1 July 1954.[17] The Joint Meteorological Committee met on 15 September 1953 and, having made some changes, approved the plans.[18] In October Commander Daniel F. Rex of the Office of Research and Invention wrote[19]

> The Joint Meteorological Committee of the Joint Chiefs of Staff has approved a plan for the establishment of a Joint Numerical Weather Prediction Unit on July 1, 1954. The mission of this unit is to provide routine operational forecasts of the 3-dimensional distribution of the meteorological elements by means of numerical computational techniques.

That it was the work at Princeton that led to the formation and funding of the JNWP Unit was stated by Harry Wexler:[20]

> With the help of these computing tools, it was found [by von Neumann's meteorology project] that forecasts over periods from 24 to 48 hours are possible, and give significant improvements over the normal, subjective method of forecasting. . . . On the basis of the results cited it was determined by the sponsoring agencies that a routine 24-36-hour numerical forecasting service . . . should be set up on a permanent basis.

In 1946, as we saw at the beginning of the last chapter, most meteorologists regarded numerical forecasting either as a complete impossibility or as a possibility only in the very distant future. As late as 1952 C.K.M. Douglas said, "On the whole the prospects of computing the future weather, with or without electronic machines, look remote at present" (p. 18). Yet by 1957 numerical forecasts were in several countries being produced daily and with success, Rossby calling them "somewhat more reliable than conventional forecasts" (1957, p. 32). How the work at Princeton and a few other places had so swift an effect on meteorological practice worldwide is the next question to consider.

The Establishment of Numerical Forecasting

There were two components of the success of numerical forecasting: an accuracy exceeding that of subjective forecasting and the capability of making the forecasts rapidly enough to be useful. The accuracy of numerical forecasting was shown in many trials made in the early and mid 1950s: most meteorologists were surprised how good the very first computer forecasts—the barotropic forecasts made on the ENIAC in 1950—proved to be, and already in 1952, with the baroclinic forecasts made on the Institute for Advanced Study Computer, one could say that numerical forecasts were on occasion better than subjective forecasts. It took only a bit longer to show that numerical forecasting was operationally feasible.

This was convincingly demonstrated in the first years of the Joint Numerical Weather Prediction Unit. As we have seen, it was mainly the results achieved in von Neumann's meteorology project that convinced decision-makers of the Weather Bureau, Air Weather Service, and Naval Aerological Service to provide for operational computer-forecasting. The JNWP Unit came into existence on 1 July 1954 and began real-time daily forecasting on 6 May 1955. But the new era in meteorological practice, whose beginning is marked by the establishment of the JNWP Unit, did not arrive unheralded: there were in the early 1950s two important trials of real-time numerical forecasting.

In 1952 at the Air Force Cambridge Research Center, Philip Thompson, Louis Berkofsky, and others, who did not at the time have access to an electronic computer, used punched-card machines to prepare 24-hour trial forecasts "under simulated routine operating conditions" (Berkofsky, 1952, p. 271). The computations, based on a linearized two-dimensional quasi-geostrophic model, made use of tabulating machines (including a multiplying punch) and of a collection of previously punched decks of cards. The forecasts, which were said to be roughly equivalent in accuracy to subjective forecasts made by an experienced forecaster, took about $11\frac{1}{2}$ h to prepare. Hence the method was too slow to be useful, but Berkofsky pointed out that if the computing were done by electronic computer rather than punched-card machines the computing time could be cut from almost 6 h to about a quarter of an hour, so "it appears possible to produce a 24-hour machine forecast in a total of about 7 hours, which is quite comparable with the time required to prepare a forecast of the 500-millibar surface by subjective methods" (Berkofsky, 1952, p. 273). This may be taken as the first demonstration of the feasibility of operational forecasting by means of computer, even though only tabulating machines were used.

The next real-time trial of numerical forecasting, and the first to use an electronic computer, took place just over a year later. In 1951 at the Institute of Meteorology of the University of Stockholm, which Rossby had recently enlarged and reorganized, a group of meteorologists had been assembled to work toward numerical forecasting.[21] The Swedish Computer Board had given these meteorologists access to BESK, which was the first large Swedish electronic computer, similar in design to the IAS Computer. Using, with some modifications, a barotropic model Charney and Phillips had devised for von Neumann's project, the Stockholm group made twenty-four 24-h forecasts in late 1953 and early 1954. For these forecasts the correlation coefficient between predicted and observed values of the height of the 500-mb surface averaged 0.77, which was somewhat better than earlier results.[22] The participating meteorologists, while admitting that a baroclinic model was on occasion necessary, felt that the barotropic model showed great promise.

In several cases the Stockholm group made the forecasts as rapidly as possible after the observations were taken "in order to gain experience for routine forecasting by numerical methods" (Staff Members, 1954, p. 141). It was found that the process took $6\frac{1}{2}$ hours: the initial data-handling (checking, plotting, analyzing, and punching onto the input tape) took 5 h 20 min, the computing took 40 min, and the plotting and analyzing of the forecast took 30 min. Since several additional hours were required for assembling the reports of observations and for disseminating the forecast, the entire process was not quite rapid enough to be useful for a 24-h forecast. The group concluded that

> . . . the time given above is not prohibitive if one can extend the forecast range to 48 or 72 hours but as soon as we want to extend the computations to two or several layers in the atmosphere other schemes for preparation of the data have to be devised. Furthermore certain errors were still found in the initial values in spite of the checking routine which was adopted here.

Thus the Stockholm group demonstrated that operational forecasting was on the verge of being feasible. They considered it especially important that a method of automatic data analysis be devised, both to eliminate errors and to reduce the time given to the initial data handling. Even without automatic data analysis, computer forecasting would, they showed, be practical if the forecast range could be extended while retaining an acceptable level of accuracy. This seemed likely, partly because earlier trials of the barotropic model had given good results with longer-range forecasts and partly because the Stockholm group saw ways of improving the forecasting procedure significantly. Their principal recommendations were that the forecast region be enlarged, that the previous forecast be used to supplement the data for regions where few observations were taken (as over the

Atlantic), and that the vorticity field be artificially smoothed to compensate for the inability of the model to compute accurately the motions on a scale only a few times the grid size. Their report, published in *Tellus* in mid 1954, was widely noticed and generated interest in putting computer forecasting into daily practice.

What made the greatest impression on the meteorological community was the establishment of numerical forecasting on a real-time, daily, and permanent basis in May 1955 when the JNWP Unit began full operation. In August of that year Harry Wexler wrote, "It has now been making daily forecasts for over 3 months, and with very good success."[23] And somewhat later, members of the JNWP Unit reported, "After almost a year of experimentation and operational numerical weather forecasting, it is concluded that the quality of the numerical 500 millibar forecasts is not significantly different from that of the best subjective forecasts prepared by methods in current use" (Staff Members, Joint Numerical Weather Prediction Unit, 1957, p. 315).

Already by late 1955 there was intense and widespread interest in numerical forecasting. A survey conducted by George Platzman of the University of Chicago reveals the extent to which dynamical methods of forecasting were used at that time.[24] In late 1955 there were only two places where an electronic computer was being used in operational forecasting: Suitland, Maryland, where George Cressman was directing the work of the JNWP Unit of the Weather Bureau, and Stockholm, where Carl-Gustaf Rossby and Bert Bolin were directing the forecasting which the Institute of Meteorology was doing in cooperation with the Swedish Air Force Weather Service. Experimental forecasting by means of computer was, however, being done in quite a few places. In the United States, besides by the JNWP Unit, experimental forecasting was being done by the General Circulation Research Section of the Weather Bureau, also in Suitland, Maryland; by the meteorology project of the Institute for Advanced Study; by the Atmospheric Analysis Laboratory of the Geophysics Research Directorate in Cambridge, Massachusetts; by the Department of Meteorology of the University of Chicago; and by the Department of Meteorology of the University of California at Los Angeles. Outside the United States experimental forecasting by means of computer was being done, besides at Stockholm, by the Royal Meteorological Institute of Belgium, by the Deutscher Wetterdienst, and by the British Meteorological Office.[25] Although not included in the Platzman survey, similar research was being done in Russia. Meteorologists at the Central Institute of Forecasting in Moscow began making trial forecasts in 1954. They used the Soviet BESM computer, comparable to the IAS Computer, and a barotropic model that was similar to ones being used in the United States and Sweden (Phillips *et al.*, 1960, p. 607).

In addition to those who were already using computers to study the atmosphere, there were in late 1955 a number of meteorologists with definite plans to begin. These included groups at the Meteorological Service of Canada, the Swedish Meteorological and Hydrographical Institute, the Norwegian Meteorological Institute, the Geophysical Institute in Bergen, the Meteorological Service of Israel,

Imperial College in London, Woods Hole Oceanographic Institution in Massachusetts, and the Department of Meteorology at MIT.[26] Some of these groups were being held back by the difficulty in gaining access to a computer.[27] The development of numerical forecasting in the Soviet Union in the 1950s lagged somewhat behind that in Western Europe and North America, partly because of a lag in computer technology.[28]

Thus in late 1955 the operational feasibility of numerical weather prediction had been demonstrated, and groups in about a dozen countries were experimenting with the method, yet in only two places were numerical forecasts being routinely prepared. A decade later at least 14 countries had operational numerical prediction: Great Britain, Norway, Sweden, West Germany, Holland, Belgium, France, Italy, Rumania, the Soviet Union, Israel, Japan, Canada, and the United States.[29] And the growth of numerical meteorology was in fact greater than suggested by the increase in the number of weather services preparing numerical forecasts, as numerical methods came to be used more and more frequently within each weather service. A measure of the internal growth is provided by the number of people in the numerical meteorology group of the Navy's Fleet Numerical Weather Central at Monterey, California: 6 in 1959, 57 in 1963, 156 in 1971, and 284 in 1979 (Bates and Fuller, 1986, p. 311).[30]

A large part of the process of generating forecasts came to be carried out by computer. In the first years of the JNWP Unit the computer forecasts were used only as advisory material by the human forecasters of the Weather Bureau, but in 1961 it became clear that for the 500-mb forecasts a computer alone did better than a human using a computer forecast as a starting point, and from then on the 500-mb forecasts were produced entirely by computer (Shuman, 1963, p. 213).[31] There remained a substantial human input in most types of forecasting, particularly in the prediction of local storms. In the 1950s and 1960s some meteorologists hoped, and many feared, that computers would replace human forecasters; there were even trials of forecasts worded by computer (see Cressman, 1972, p. 188). But as it became clear that certain human skills could not soon, if ever, be set aside, the main issue of discussion became not whether computer forecasting was better than human forecasting, but what the optimal man-machine combination was. Accordingly, much research has been directed toward facilitating interactive forecasting, especially through the use of computer graphics (Snellman and Murphy, 1979, pp. 800–803).[32]

In reviewing the developments of the first postwar decade, we must bear in mind that at the end of the war there was in the United States only a single electronic computer, and its capacities, as we saw in Chapter 10, were so limited that they allowed only a greatly simplified form of numerical forecasting and were of little help in dealing with "data push." So we turn now to the development of the technology that not only made possible effective numerical forecasting and effective processing of observational data, but also led to a new style of meteorological research and to a unification of the discipline.

The Computers Used by Meteorologists

Meteorologists were among the users of many of the first generation of computers.[33] The Princeton group used, of course, the IAS Computer, a machine that did much to set standards for later machines. The two groups in Suitland, Maryland, were using an IBM 701. The 701, which was introduced in 1953 and was the first commercial electronic computer made by IBM, was similar in design to the IAS Computer but used card readers and card punches for input and output. As already remarked, the groups at Stockholm and at Moscow were using roughly comparable machines, the BESK and the BESM, respectively. The Belgian meteorologists used a less advanced computer belonging to the Institut de la Recherche Scientifique at Antwerp. The group at Los Angeles used the SWAC, which from mid 1950 until the completion of the IAS Computer in mid 1951 was the fastest computer in the world. The English meteorologists had access to the computer at Manchester University, the Ferranti Mark I, the prototype of which has a claim to the honor of being the first functional stored-program computer. And the group at MIT used the Whirlwind computer, which pioneered new technology in a number of areas, including graphic display of output and magnetic-core memory (which is discussed below).[34]

The spectacular improvements in speed, memory capacity, and reliability of electronic computers in the 1950s and 1960s were crucial to the widespread adoption of the computer as a meteorological tool.[35] The IBM 704, introduced in 1955, was about five times as fast as the 701; a decade later the fastest of the computers being marketed was about a hundred times as fast as the 704 and performed almost a million arithmetical operations per second. When the JNWP Unit acquired an IBM 704 in 1957, it could then implement more sophisticated models (Thompson, 1986, p. 31).[36] Whereas the cost of a large computer system of the sort used for operational forecasting remained roughly constant from 1955 to 1967, the cost of 5 million computations fell from $42.00 to 20¢ as a result of the increase in speed.[37]

The limited memory capacity of early computers was troublesome to most users, but was especially troublesome to meteorologists, who typically had to deal with both complex programs and many data. In Chapter 10 we saw the inconvenience caused by the fact that ENIAC could store only twenty 10-digit numbers. The BESK computer used by the Stockholm group had a memory of only 512 words of 40 binary digits each. Thus after these meteorologists had read in the data representing the current state of the atmosphere, husbanding space by putting the three quantities associated with each grid point in a single word, only 112 words remained for the instructions. It was therefore necessary to read instructions in (from paper tape) continuously during the computation. Even though the internal memory of the Institute for Advanced Study Computer was twice as large as that of BESK, and the internal memory of the IBM 701, with which the JNWP Unit began operations in 1955, was almost twice again as large (2048 words of 36 binary digits each), limited memory capacity continued to cause major difficulties for meteorologists.

The basic element of the internal memory of many of the early electronic computers, such as the EDSAC, the UNIVAC, and the Pilot ACE, was the acoustic delay line, which stored information in the form of sound waves in a tube of mercury or other substance. In the late 1940s F.C. Williams of the University of Manchester invented a storage device that incorporated a cathode-ray tube. The Williams tube, although still bulky and expensive, was somewhat more reliable and much faster than a delay line. The computers mentioned in the preceding paragraph—the BESK, the IAS Computer, and the IBM 701—all used Williams tubes for internal storage. In 1952 a new, much cheaper, and much more reliable device, the magnetic core memory, was demonstrated on a test machine at MIT. According to Maurice Wilkes, designer of the EDSAC, this invention was "the one single development that put computers on their feet" (Wilkes, 1985, p. 209). The difference in speed and reliability between core memory and Williams tube memory was shown by the experience with Whirlwind computer at MIT, one of whose users was V.P. Starr of the Department of Meteorology. When Whirlwind was converted to core memory its operating speed doubled, its rate of inputting data quadrupled, and the maintenance time on its memory fell from 4 hours per day to 2 hours per week.[38] The first IBM computers with magnetic core memory were the 704 and the 705, both announced in 1954. Initially the 704 had a memory of 4096 words; in early 1957, IBM 704s with 32,768-word memories were delivered (Bashe *et al.,* 1986, p. 180).

In the early 1960s the limited size of internal memory was still a major problem. Since the grid of a forecasting algorithm had 2000 to 16,000 points, with the values of several meteorological variables specified at each point, it still was not possible to read in all of the data at one time. This meant that computations had to be interrupted repeatedly to read data in and out (Wolff and Hubert, 1964, p. 640). As the size of the internal memory increased in the late 1960s this problem was alleviated.

There were important gains in reliability and speed when, in the late 1950s, transistors began to replace vacuum tubes on a large scale. One of the first computers to be designed for transistor electronics was the IBM Stretch computer (also known as the IBM 7030). It was completed in 1961, and the following year one was put into service by the General Circulation Research Laboratory of the Weather Bureau.[39] Transistor electronics and a sophisticated look-ahead scheme of instruction processing made Stretch almost a hundred times as fast as the IBM 704 at adding and multiplying (Williams, 1985, pp. 393–394).

In the 1960s a great variety of large, sophisticated computers became available. In 1964 IBM launched its 360 line of computers, which came to be used by quite a few meteorological groups. The acquisition of a 360/65 by the Canadian Meteorological Centre in 1967 permitted the use of more sophisticated forecasting algorithms, and the British Meteorological Office had to wait until the delivery of its 360/195 to attempt quantitative forecasts of rainfall by numerical means (Kwizak, 1972, p. 1155; Mason, 1970, p. 28).[40] Control Data Corporation successfully marketed its 6000 series; the Weather Bureau acquired a CDC 6600 in 1965, and the Fleet Numerical Weather Central, a CDC 6500 in the following year. According to Philip Thompson, "about the time the CDC 6600 came out [in 1965] there was a more or less abrupt increase in the order of complexity that could be handled on the machine" (1986, p. 38). The increased computational power permitted more detailed treatment of the phenomena included in models of the atmosphere; Thompson cited the turbulent transfer of momentum in the boundary layer of the atmosphere as a phenomenon that therefore could be handled in a more satisfactory way. The phenomenal increase in computing speed meant that by the mid 1960s many meteorological tasks could be effectively carried out by medium-priced computers (Wolff and Hubert, 1964, p. 640), and the so-called "minicomputers," introduced by Digital Equipment Corporation in 1963, became popular with meteorologists (and scientists generally). By 1970 computers using integrated circuits (entire circuits on silicon chips) were being marketed; such computers were more powerful, smaller, and much faster than earlier computers.

The importance of computing power to U.S. weather services in the late 1960s is indicated by a list of the large computers located at the three largest meteorological centers in 1969: the Weather Bureau at Suitland, Maryland, had two CDC 6600s; the Air Weather Service at Offutt Air Force Base in Nebraska had four Univac 1108s; and the Fleet Numerical Weather Central in Monterey, California, had two CDC 6500s, two CDC 1604s, two CDC 3200s, and one CDC 8090 (Bates and Fuller, 1986, p. 189). In the 1970s and 1980s advances in computer technology continued to affect meteorology significantly. In 1976 the first of the Cray computers became available, and many weather services have since acquired Cray computers. The rapid adoption of personal computers in the late 1970s brought, according to Richard Heim, "dramatic changes to the fields of meteorology and climatology, especially with regard to research, educational, and operational activities" (1988, p. 490).[41]

The result of this technological development was that many of the constraints on numerical forecasting were relaxed. By the early 1960s meteorologists were much less troubled by computer unreliability. In 1964 Navy meteorologists reported, "Since the 1961 generation, all computers have had acceptable reliability except for tape operations," and "Scheduled maintenance time has been held to less than two hours per day" (Wolff and Hubert, 1964, p. 641).[42] Limited memory-capacity likewise became less of a problem. Computational speed, despite the great increases of the 1950s and 1960s, came to be the major constraint imposed by the hardware. Thus meteorologists vigorously sought the most powerful computers available, and many meteorologists came to feel what Akira Kasahara expressed in 1987, "When it comes to the bottom line, whoever has the faster computer will win" (Gleick, 1987b).[43]

Meteorology has had some influence on the development of computer technology. From the 1940s to the present, meteorology has been an eager consumer of a wide range of computer products and services, in particular of the most advanced machines. So even though the number of meteorologists has been relatively small, the requirements of numerical weather forecasting have on occasion been considered by those designing hardware (Pugh, 1984, p. 161).[44] A few of the designers of computers were themselves interested in certain parts of meteorology, notably John Mauchly, John von Neumann, and Maurice Wilkes.[45]

Some mention must be made of the development in the 1950s and 1960s of what came to be called computer software. In the early years of numerical meteorology, computers were programmed in so-called machine language, with instructions and memory locations referred to by numbers, often hexadecimal or octal numbers. In the early 1950s so-called assembly programs began to be used; they eased the programmer's labor by allowing the use of mnemonic codes for instructions and "relative addresses" for memory locations (with the assembly program making the actual assignment of locations). Because the instructions were very simple ones—the basic arithmetical operations (and sometimes not division) and simple logical operations such as testing if a number is positive—and because internal memory was quite limited, programming remained an extremely challenging task.[46]

Beginning in the late 1950s higher-level programming languages such as Fortran made it much easier for meteorologists to use computers. These languages, unlike the machine and assembly languages, were not machine specific, which meant that the sharing of programs also became much easier. For the most demanding computations, however, Fortran and the like could not be used because greater speed and more efficient use of the limited core storage could be achieved using a lower-level language (Wolff and Hubert, 1964, p. 641). The fact that meteorologists often wanted to use computer speed and memory to the utmost also meant that company-provided software packages often could not be used (Wolff and Hubert, 1964, p. 641). Thus meteorologists have themselves been very active in the development of sophisticated software, notably software for parallel processing.

The Abandonment of Other Calculating Aids

As computers proliferated on the meteorological scene, they pushed most other calculating aids off the stage altogether. The oldest of the calculating aids, numerical tables, managed, but just barely, to maintain a presence.[47] For example, in 1949 J.C. Thompson published a set of tables for computing the heights of the 850, 700, and 500-mb surfaces from the readings of altimeters in aircraft. In each of the early volumes of *Tellus,* the journal Rossby established in 1949, there are several articles that include tables for calculating, and this is true as well of the contemporary volumes of the *Bulletin of the American Meteorological Society.* In 1966 the U.S. Weather Bureau published a collection of tables to be used in the processing of radiosonde data. But in the 1960s and 1970s electronic computers and hand-held calculators came to replace tables for routine or complicated tasks, although tables continued to be used occasionally.[48] It must be remembered that until quite recently computers were very expensive, so many meteorologists, even in industrial countries, had to make do with tables and other inexpensive calculating aids.

Geometrical methods of calculation, that is, ones that involve the representation of a quantity by a length or an area, have been abandoned almost entirely. It was in the 1940s and 1950s that interest in special-purpose slide rules, nomograms, special instruments and paper for plotting data, and other graphical techniques was at its highest, and interest remained high in the 1960s. In 1949 O.G. Sutton described a slide rule to be used for calculating atmospheric diffusion, and in the 1950s several different circular slide rules were constructed for calculating the winds in the flight path of an aircraft (Buell, 1961). In 1959 Frank Hanson and Paul Taft devised a plotting system, involving several movable scales and a two-piece plotting surface, for finding wind speeds by tracking pilot balloons using two theodolites; Hanson and Taft devised the system in order to "eliminate the laborious process of hand computing double-theodolite balloon-measured wind data" (1959, p. 221). Two years later Robert Weedfall and Walter Jagodzinski (1961) presented a different graphical procedure for the same task. Because the calculations to be carried out with the raw data often depended on particularities of the observational instrument, an appropriate slide-rule, chart, or plotting form was in many cases provided with the instrument (see, for example, MacCready, 1957).

Calculations involving graphs, either on maps or on more abstract coordinate planes, were much used. In such calculations integrations were sometimes done using planimeters (Robertson and Cameron, 1952). In 1950 A. F. Williams showed a way to use a map as a calculating aid: if barometric pressure is plotted on map for which the scale at each point is proportional to the sine of the latitude, then the geostrophic wind speed at any point is proportional to the contour spacing there. Another example is Harold Crutcher's (1956) graphical procedure for estimating the effect of wind on aircraft speed. And because statistical procedures frequently involved a great deal of computation, graphical procedures were sometimes used for their facilitation (Godske *et al.,* 1957, pp. 750–751, 763–765).

In the 1950s and 1960s nomography was perhaps the most commonly used of the graphical methods. Nomograms were constructed for use in predicting the location of the jet stream, for estimating the return periods of extreme weather conditions, for finding the amount of precipitable water, for determining the floating altitude of balloons, and for dozens of other meteorological tasks.[49] A nomogram was generally easy to use, the only physical operation being the alignment of a straightedge, and it could obviate a complicated and time-consuming analytical or numerical procedure. A particularly intricate example, constructed by Frank Gifford in 1952 to calculate atmospheric diffusion, is shown in Figure 1. Whereas most nomograms were made to solve one equation, this nomogram, which comprises thirteen scales, solves nine different equations (Gifford, 1953, p. 103).

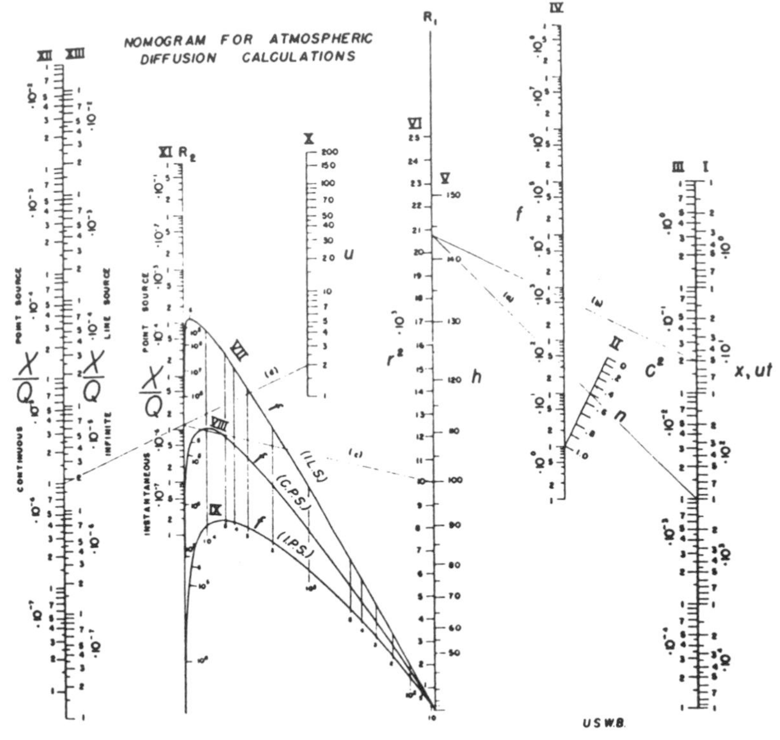

Figure 1 This nomogram, devised by Frank Gifford, allows one to solve quickly nine equations which O. G. Sutton had derived from describing atmospheric diffusion (© American Meteorological Society. From Gifford, 1953).

Nomograms, like other analog devices, had quite limited accuracy. The above nomogram produced only one or two significant figures. According to the inventor, "This is no drawback, since, at present, such accuracy considerably exceeds that of both diffusion theory and experimental concentration measurements" (Gifford, 1953, p. 105). Yet as theory and observation advanced, there came to be fewer and fewer areas where one-place accuracy was satisfactory. The 1960s were the last decade in which an appreciable number of new nomograms were reported in the meteorological literature.[50]

One of the most important graphical techniques of the 1950s was the method devised by Ragnar Fjörtoft (1952) for integrating the barotropic vorticity equation. As described in Chapter 8, Fjörtoft was working in the Norwegian tradition of the graphical calculus, and his 1952 paper may be said to have fulfilled Bjerknes's program of calculating the weather graphically. The Platzman survey shows that in late 1955 Fjörtoft's method was being used in quite a few places: in Tokyo by both the Central Meteorological Observatory and the Geophysical Institute of Tokyo University, in Dublin by the Irish Meteorological Service, in Wellington by the New Zealand Meteorological Service, and in the United States by seven groups.[51] In about 1951 the Soviet meteorologists N.I. Buleyev and M.I. Yudin devised graphical techniques, similar to Fjörtoft's, for solving the vorticity equation and related equations, and their techniques were used in trial forecasts (Phillips *et al.,* 1960, pp. 604–605). Yet as computers became available the graphical calculus dropped out of use.

In meteorology, as elsewhere in the scientific and business worlds, the tabulating machines for which punched cards were originally designed were gradually replaced in the 1950s and 1960s by electronic computers. Punched cards, however, came to be even more common, both for storing data and for reading data in to a computer. As we saw in Chapters 8 and 9, by the 1950s meteorological data were, in many countries, routinely stored on punched cards. In 1956 a committee appointed by the World Meteorological Organization to plan the International Geophysical Year recommended that the practice be more universal:

> All meteorological services which can do so should prepare their own punch cards of marine data. So far as is known at present, the following countries will tabulate their own cards; Federal Republic of Germany, France, Japan, Netherlands, United Kingdom, U.S.S.R. and U.S.A. Other countries punching cards may either tabulate their own cards or make a bilateral arrangement for their tabulation by one of the above countries or by any other country using mechanical tabulation methods (p. 105).

In the 1950s the British Meteorological Office made even greater use of punched cards than it had earlier, for example in computing linear-function tables for upper-air data and in preparing climatological atlases of the oceans (Casey *et al.,* 1958, p. 319). In 1955 C.L. Godske of the Geophysical Institute in Bergen wrote[52]

> I have planned to "translate" the graphical method of Fjörtoft and perhaps also to modify it, so that it can be carried through by means of punch card methods, punch card machines being the only ones available at present at our institute—and probably also for many years to come.

The Upper Atmosphere Research Laboratory of Boston University made bibliographic use of IBM cards in the early 1950s. For each article to be included in the bibliography, information about author, journal, date of publication, subject area, language, and security classification was punched on a card, and on the back of the card an abstract of the article was typed. With the use of a sorter, such as an IBM Type 083, one could quickly obtain useful subsets of the bibliography (Casey *et al.,* 1958, p. 318).[53]

Another type of calculating aid that became obsolete in the 1950s and 1960s was analog computers. Kelvin's tide predictor, Bush's differential analyzer, and the linear-equation solver of RCA, all mentioned in previous chapters, were analog computers which, although built for other purposes, were applied to meteorological problems. It was especially differential analyzers that meteorologists found useful; they were used to solve a nonlinear system of equations advanced by Edward Lorenz and to find the spectral representation of functions as suggested by George Platzman; Maurice Wilkes modeled atmospheric tides with the differential analyzer at the Mathematical Laboratory of Cambridge University; and Luigi Jacchia and Zdenek Kopal investigated atmospheric oscillations and temperature profiles with the Rockefeller Differential Analyzer at MIT (Dingle and Young, 1965, p. vi; Wilkes, 1949; Jacchia and Kopal, 1952). We saw in Chapter 8 that Mauchly designed an analog computer for a particular meteorological application, and in the 1950s quite a few meteorologists followed his example. R.J. Taylor and E.K. Webb (1955), finding that a study of the vertical turbulent fluxes of heat, water vapor, and momentum involved a great deal of calculation with the raw data, designed a special-purpose differential analyzer for this work.[54] D.P. Brown and R.A. Harvey (1961) built an integrator to record automatically a weighted average daily solar radiation. Seymour Hess (1957) designed an analog computer for finding the Laplacian of any mapped quantity, such as the geostrophic vorticity. And in the late 1950s the Central Institute of Forecasting in Moscow used a special-purpose electronic analog computer for predicting temperature anomalies (Baum and Thompson, 1959, p. 405).

With all these devices, accuracy was quite limited, usually to one or two significant figures. When it was used in a meteorological study in 1946 the RCA linear-equation solver was judged unsatisfactory because of its limited accuracy.[55] In the study made by Jacchia and Kopal referred to above, the differential analyzer was abandoned in favor of mechanical desk calculators whenever greater accuracy was required. It was sometimes possible, by reformulating the mathematical task, to increase somewhat the accuracy achievable,[56] but limited accuracy remained a problem, which, together with the fact that most analog devices could solve only a narrow range of problems, led to the dominance of the digital computer in the 1960s.

Having looked at the revolution in computing technology, we turn now to a brief review of the development of forecasting algorithms and general-circulation modeling that in the late 1950s and early 1960s secured the acceptance of numerical meteorology.

Advances in Numerical Meteorology

The beginnings of computer forecasting were recounted in Chapter 10: the 1-dimensional model based on vorticity conservation, which was used in hand computation in 1949; the 2-dimensional barotropic model, which was run on the ENIAC in 1950; and the $2\frac{1}{2}$-dimensional baroclinic model, which was run on the IAS Computer in 1952. This baroclinic model divided the atmosphere into two layers, and researchers at Princeton and elsewhere soon increased the number of layers, which permitted more accurate modeling of vertical motion and density stratification (Thompson, 1978b, p. 145).

Another trend was that models came to incorporate more and more phenomena, such as improved wind approximations, effects of surface friction, representation of mountains, latent heat of condensation, and exchange of heat between the surface of the earth and the atmosphere (Cressman, 1972, p. 187). The first attempt to predict precipitation by numerical means was made by Joseph Smagorinsky and G.O. Collins in 1955. This was an especially difficult task because the physics of cloud formation and precipitation was only poorly understood and because the convective motions that give rise to much of the precipitation occur on a smaller scale than is resolvable in the usual numerical models. As a result it was not until the mid 1960s that the transformations of water vapor were regularly included in numerical models (Smagorinsky, 1970, p. 28). In about 1960 the effects of mountains and friction were added to the model used by the JNWP Unit. This enhancement of the model corrected its earlier tendency to forecast the position of the Aleutian Low too far eastward (Shuman, 1963, p. 213).

A momentous development of the late 1950s was the implementation of models based on the so-called primitive equations, the basic equations of physics describing the atmosphere. These were the equations Bjerknes singled out at the turn of the century and the ones Richardson integrated numerically. Some work by Charney in 1955 indicated that the use of the primitive equations might be made practical, but it was not until the mid 1960s that, through the earlier efforts of Arnt Eliassen, Joseph Smagorinsky, K.H. Hinkelmann, F.G. Shuman, and others, primitive-equation models replaced baroclinic models for daily forecasting (Cressman, 1972, pp. 184, 187). A prerequisite of the change was the greater speed of computers; the fact that the primitive-equation models included gravity waves meant that, in order to avoid computational instability, the time step had to be shortened from 1 hour for the earlier models to 10 minutes (Shuman, 1963, p. 214). A result of the change was that the very latest forecast algorithms then resembled the very first. In 1967 George Platzman wrote

> A comparison of Richardson's model with one now in operational use at the U.S. National Meteorological Center [which was devised by Shuman and Hovermale] shows that, if only the essential attributes of these models are considered, there is virtually no fundamental difference between them. Even the vertical and horizontal resolutions of the models are similar. (p. 514)

The improvement of forecast algorithms was closely related to the development of general-circulation models. With these models the intention was to simulate the time-averaged properties of atmospheric motion—properties such as the jet stream and the Prevailing Westerlies—in other words, to produce what von Neumann called the "infinite forecast."[57] This line of development was established in 1955, when Norman Phillips designed the first numerical model able to reproduce the principal motions of the earth's atmosphere. Phillips's model worked so well that it was taken as evidence that researchers understood the basic action of the atmosphere. Joseph Smagorinsky has written

> . . . Norman Phillips had completed, in mid-1955, his monumental general circulation experiment. . . . The enabling innovation by Phillips was to construct an energetically complete and self-sufficient two-level quasi-geostrophic model which could sustain a stable integration for the order of a month of simulated time. Despite the simplicity of the formulation of energy sources and sinks, the results were remarkable in their ability to reproduce the salient features of the general circulation. A new era had been opened. (1983, p. 25)

The publication of Phillips's model in 1956 contributed to a surge of interest in modeling the global circulation of the atmosphere. Other contributing factors were that electronic computers were newly available to meteorologists, that great advances were made in making meteorological measurements worldwide and at various heights,[58] and that there was a rapid development of meteorological theory. Interest in global circulation modeling received another boost in 1963 with the publication of Joseph Smagorinsky's primitive-equation model of the general circulation. As with forecasting, the rapid development of computers in the 1960s and 1970s was a continuing stimulus to improvement in circulation modeling.[59]

A vital characteristic of numerical forecasting and of general circulation modeling has been the possibility of fairly systematic progress. The progress has taken place, in large part, along certain well-known but often steep tracks: increase in the spatial and temporal resolution of the model, extension of the spatial and temporal range, inclusion of more atmospheric processes in the model, improvement in the way a particular process is modeled, and improvement in the numerical methods, programming, and computer hardware. And since a forecasting algorithm, unlike the skill of a human forecaster, is a communal artifact, put together of fully specified parts in a fully specified way, the improvements have been widely shared.

The best documentation of the continual, though gradual, progress since the mid 1950s is provided by measures of forecast accuracy. Although the evidence is less complete for the decades prior to 1950, there seems to have been little improvement in forecast accuracy prior to the mid 1950s.[60] Figure 2 shows average forecast error for two types of forecasts made by the Weather Bureau.

The decline in error from 1955, when computer-generated forecasts began to be used, until 1971 was fairly regular, and the improvement in the prediction of pressure has continued at about the same pace through the mid 1980s (Tribbia and Anthes, 1987, p. 495). Although the improvements in temperature and precipitation forecasts have not been as great, they have been measurable and significant (Reed, 1977; Shuman, 1978; Kerr, 1985).

The 1950s and 1960s were years of fundamental change in meteorological research and weather forecasting. At the root of this change was the new availability of electronic computers. We have seen in this chapter that it was the early success of numerical forecasting, reinforced by steady improvement in forecasting and by success in general-circulation modeling, that induced meteorologists to take up the computer. The next chapter explores the consequences of the adoption of computers.

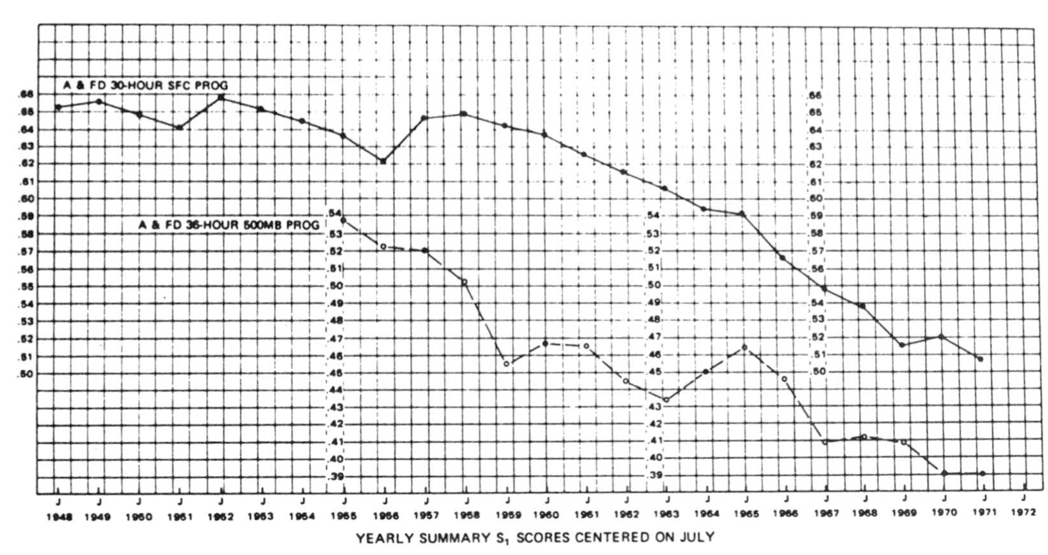

Figure 2 This graph, reproduced from Cressman (1972, p. 185), shows average error on two types of prognostic chart, the 30-h prediction of the height of the 1000-mb surface (which gives the barometric pressure at the surface) and the 36-h prediction of the height of the 500-mb surface (which gives much information about the large-scale motion of the atmosphere). Cressman writes that "Prognostic charts which appear practically perfect to the eye would score 0.30 at sea level and 0.20 at 500 mb."

Chapter 12 The Unification of Meteorology

Growth of the Profession

The 1950s and 1960s were years of steady growth for the meteorological profession. In 1945, *after* the wartime spurt, the American Meteorological Society had 2883 members; in 1955 the members numbered 5449; in 1975 8538 (Bates and Fuller, 1986, p. 298).[1] It became more common for meteorologists to be employed outside government and academia. This is reflected in three programs which the American Meteorological Society established in the 1950s: the certification of consulting meteorologists, the recognition of competence in radio and television weather forecasting by the awarding of a "Seal of Approval," and the institution of an award for excellence in applied meteorology.[2] Also notable is the fact that the number of AMS corporate members, most of them businesses, increased from 45 in 1955 to 117 twenty years later.[3]

The growth of meteorology in government and academia was even greater. In the United States, appropriations for the Weather Bureau climbed from $25 million in 1955 to $108 million in 1965 to $230 million in 1975, and expenditures for military weather services grew at roughly the same rate (Whitnah, 1961, p. 21; Bates and Fuller, 1986, p. 179). From the mid 1940s to the 1980s there was almost a 10-fold increase in the number of American universities awarding advanced degrees in meteorology (Thompson, 1987, pp. 632, 635). The American Meteorological Society published only one journal from its founding in 1919 until 1944; by 1975 it was publishing seven. The amount of research, most of it paid for by the Federal government,[4] increased spectacularly: in the four decades after World War II there was at least a 10-fold increase in the number of articles published annually in American journals of meteorology (Thompson, 1987, p. 632).

With the postwar decades came a general heightening of interest in meteorology. The surge of interest during World War II in applied meteorology continued; "engineering meteorology" and "applied climatology" became common terms as meteorological information came to be used in making decisions about plant location, vehicle design, choice of construction materials, and a multitude of other matters (Hewson, 1963). Other contributing factors were the great increase in commercial aviation, a concern about air pollution, and some success in weather modification.[5] National weather services increased their activities; there were more frequent and more detailed forecasts, and more types of specialized forecasts, such as tornado warnings, hail forecasts, and warnings of flash floods (Bigler, 1981, p. 159)

International cooperation too reached new levels, and here computers played an especially important role. Like the synoptic method of analysis and forecasting 80 years earlier, numerical weather prediction and numerical modeling of the atmosphere in the 1950s and 1960s invigorated international cooperation. Numerical models, whether for operational use or for research, typically required data from an extensive area. Global circulation models stimulated interest in the regions for which there were only sparse data, such as the tropics, the oceans, and Antarctica (Smagorinsky, 1970, p. 31). A principal aim of the International Geophysical Year, which ran for 18 months in 1957 and 1958, was the collection of meteorological data over the entire globe, with an eye to improving numerical forecasting (World Meteorological Organization, 1956, p. 7). As meteorologists worked to extend the range of forecasts beyond several days, they found that boundary effects were vitiating forecasts, hence they enlarged the areas represented in forecasting models. By 1964 most operational models used data from much of the Northern Hemisphere, and in 1968 Jerome Namias wrote, "Theory and evidence indicate that the entire global atmosphere must be considered for extended range predictions" (Wolff and Hubert, 1964, p. 640; Namias, 1968, p. 445). Two of the largest international projects in any science and in any period were motivated largely by the interest in improving computer forecasting and computer modeling; these were the World Weather Watch and the Global Atmospheric Research Program (GARP), both of which began in the 1960s (Namias, 1968, p. 445; Smagorinsky, 1970, p. 24).

Advances in Observational Techniques

All these developments were related to changes in the way in which the elements of the weather were observed and measured. One of the most important changes was the introduction of radar meteorology. Early in World War II the fact that thunderstorms cast radar images was only a problem, since these images obscured the images of aircraft. But by 1944 military meteorologists were being trained in the use of radar, and within a dozen or so years radar meteorology became a major branch of the science. This vigorous growth owed much to computers, which were used in the theoretical research that led to new types of observation, in the processing of the radar signals, and in the treatment of the data. For example, the processing of the signals for doppler-radar observations was so complex that it was not until the availability of minicomputers in the 1970s that this technique, first demonstrated in 1956, became operationally feasible (Bigler, 1981, p. 161). Radar meteorology, by providing new sorts of data, gave new life to certain branches of the science—such as cloud physics, hail research, and storm modeling. Computers were much used in the data handling and in the modeling, and these data and models contributed to the unification of the discipline. W. F. Hitschfeld wrote that from the 1950s on radar meteorology served "to link the two solitudes of meteorology—the microphysics of clouds and the continental and global patterns of the synopticians and the dynamicists" (1986, p. 36).

A second new mode of observation, meteorological satellites, also grew out of wartime technology, in this case rocketry. The postwar development of rocketry had some connection with meteorology, as in the use of the captured V-2s mentioned in Chapter 9. Just $2\frac{1}{2}$ years after the orbiting of the first artificial satellite, Sputnik I, the United States successfully launched a satellite devoted to meteorological observation, TIROS I on 1 April 1960. This came about with the strong support of Francis Reichelderfer, Chief of the Weather Bureau, and Harry Wexler, the Director of Research, and the United States maintained its lead in the deployment of meteorological satellites. In all, 10 TIROS satellites were placed in orbit. Three other series of U.S. weather satellites were started in the 1960s: the Nimbus series in 1962, the ESSA series in 1966, and the ATS series in 1966.[6]

The data provided by satellites had a great impact on meteorology. The fact that the entire earth has been photographed daily since 1966 has meant that tropical storms do not reach land without warning, and the "movies" of cloud motion have led to dramatic improvement in the prediction of local storms (Schnapf, 1977, p. 190; Smith *et al.,* 1986, p. 455). Satellites have been vital in providing data of the type and distribution required for forecasting algorithms. In the United States the connection between numerical weather prediction and satellite observation was vigorously promoted by Harry Wexler, who saw to it that the planning for meteorological satellites involved those working in numerical forecasting.[7] Many of the observation systems deployed in satellites were devised specifically to provide the data needed for forecasting algorithms (see Smith *et al.,* 1986). But the sorts of data provided by satellites often required accommodation on the part of the forecasters; for example, the fact that the satellite data were asynchronous (taken at different times for different regions) posed major problems of data analysis (Cressman, 1972, p. 191).

Meteorological use of satellites, as with radar, was closely tied to computer technology. Quite apart from their role in placing a satellite in orbit and communicating with it, computers were indispensable because of the unprecedented "data push" that satellites caused. Less than 3 years after the launching of TIROS I, Walter Marggraf wrote, "The vast amount of data made available everyday by the satellites to the meteorological community presents an almost overwhelming analysis problem" (1963, p. 1).[8] More satellites with continually improved sensing systems led to ever more data. The continued force of this data flow was stated by W. L. Smith *et al.* in 1986: "Management of this vast quantity of data will be among the greatest challenges for the scientific use of satellite observations of the future" (p. 462).

In the 1950s and 1960s the swelling of the flood of data—especially through the use of radar and satellites—made the rapidity of data processing more important than ever before. For climatologists, who had long had more data than they could conveniently handle, it was a familiar, although unusually pressing, situation. For forecasters the new data push was more unsettling, since for them information about the current state of the atmosphere was useful only if processed within a few hours.[9] Thus the new data gave enormous impetus to a movement long under way, to automate data handling.

The desire to reduce costs, to make data processing more objective, to reduce errors, and to deal with the large amounts of data all contributed to the movement to automate data handling. Especially important were the last two factors, as is suggested by a statement made in a 1953 proposal for a study project at the U.S. Weather Bureau:[10]

> In analyzing the meteorological data problem in general, one finds that in addition to the unnecessarily large time-lag between the taking of an observation and its ultimate usable form, there are many opportunities for errors to be introduced. In both instances the difficulty can be traced to the human element.

The proposal recommends the study of automatic observing devices, automatic data transmission, and automatic data analysis for the reason that "Human intervention should be avoided whenever possible." In 1961 Philip Thompson wrote, ". . . the efficiency of the whole data-processing chain is not much greater than that of its least efficient link. Aside from the question of human error, it will be necessary to make the entire system automatic on grounds of time economy alone" (p. 166).

Already by the late 1950s computers were doing much to relieve the burden of data handling. In 1957 Carl-Gustaf Rossby wrote

> In Washington the teleprinter tapes which contain the meteorological observations are introduced directly into the electronic computer which sorts the data wanted, checks them for internal consistency, discards faulty observations, and then stores approved observations in the machine for final analysis. (p. 37)

By the mid 1960s the task of input data handling was fully automated at the National Meteorological Center of the Weather Bureau; only about 10 min were required to process some 2000 separate surface and upper-air reports (Cressman, 1965, p. 325). The detection of bogus data had always been an urgent task for meteorologists, and here computers proved especially effective (Gustavsson, 1981, p. 73).[11] This has continued to be a vital task: in 1980 some 5 or 10% of surface-based observations contained large errors that had to be detected during processing (Gustavsson, 1981, p. 67).[12]

The next task, known as synoptic analysis, was to derive from the current observations (and sometimes also from past observations or forecasts) the values of the meteorological variables at the regularly spaced grid points of the numerical model. During the first half-year of forecasting by the JNWP Unit, the synoptic analysis was done by hand. Then an objective analysis program, written by A.L. Stickles and George Cressman, began to be used, although there was still some human intervention in the preparation of the synoptic charts (Staff Members JNWPU, 1957, pp. 265–266). Hand analysis was dispensed with entirely by 1958

(Shuman, 1963, p. 213). Similar steps were being taken in Stockholm. In 1957 Rossby reported

> Very great progress has been made lately in the problem of letting electronic computers take over the very laborious job of analyzing daily synoptic maps. The 500-mb maps, which form the basis for the numerical forecasts in Stockholm, are being analyzed by the electronic computer in such a way that the preceding day's 24-hour prediction is selected as the start for and the first approximation to the day's map. The approximation is then corrected by means of observations from the day in question. (p. 37)

Finally, the data, when they were not used exclusively as computer input, had to be transformed so that they could be readily interpreted by human forecasters. To this end the automatic drawing of contour charts began at the Weather Bureau in 1955, with significant improvements being made in 1960 [Bedient and Neilon, 1961?; *BAMS* **42,** 214 (1961)]. In 1964 Jacob Bjerknes wrote, "Electronic automation has already relieved him [the forecaster] of much of the overwhelming load of data handling, and now also presents him with electronically computed forecast maps." (p. 314).

Thus in the postwar decades steady increase in support for meteorology, especially from governments, led to phenomenal advances in observational meteorology, and "data push" and the success of numerical weather prediction led to the use of computers by many national weather services. It was not only meteorological observation and weather forecasting that were transformed.

A New Style of Research

With the introduction of the computer into meteorology came a new style of research in which numerical modeling was the dominant methodology. By the 20th century physicists and meteorologists had devised mathematical theories for many of the individual processes that take place in the atmosphere, but the mathematical intractability of many of the theories and the complexity of the action of the atmosphere—usually involving many different processes—made quantitative investigation in most cases impracticable. The computational power of the computer, together with the appropriate numerical methods, rather suddenly changed the situation. In 1952 George Platzman wrote

> . . . consider the problem of forecasting radiational temperature changes in the free atmosphere. A great deal is known about the theory of such changes: they are controlled by the distribution of water vapor, carbon dioxide, ozone, and other principal absorbers of long- and short-wave radiation, and by the distribution of liquid water (clouds), as well as the reflectivity and temperature of the underlying ground surface. Numerous graphical, tabular, and other approximate methods have been devised to account for most of these factors, but a more thorough treatment of the problem probably rests on detailed integration of the complete radiational transfer equations.

> The complexity of these equations is staggering. . . . . . However, it seems likely that a great deal could be learned by using a high-speed computer to integrate the transfer equations under conditions somewhat more complex than can reasonably be treated by means of "hand" computations. (p. 177)

A few years later Jule Charney wrote

> For the first time the meteorologist possesses a mathematical apparatus capable of dealing with the large number of parameters required for determining the state of the atmosphere and of solving the nonlinear equations governing its motion. Earlier, for want of such an apparatus, both the forecaster and the investigator were forced to content themselves with such highly oversimplified descriptions or models of the atmosphere that forecasting remained largely a matter of extrapolation and memory, and dynamical meteorology a field in which belief in a theory was often more a matter of faith than of experience. (in Aspray, 1990, p. 152)

Thus the computer was quickly adopted as a tool to investigate mathematical theories. Soon, however, the numerical models themselves came to be seen as end products of meteorological research, and these products, although logically independent of any calculating aid, were so complex that the electronic computer was an essential medium for their expression.

We have already seen how many meteorologists turned their attention to forecast algorithms and global circulation models. Also in the 1950s numerical models of individual processes—such as the growth of cloud droplets, the separation of electrical charge in cumulus clouds, and the intensification of winds in a hurricane—became common. The aim of more and more meteorological research came to be the construction of such models. In 1970 B. J. Mason wrote

> Apart from the question of atmospheric predictability, the basic scientific problem is to formulate realistic physico-mathematical models of the atmosphere that will adequately represent the complex physical and dynamical processes that are likely to control developments on the appropriate time scale (p. 30)

Numerical simulation provides a test of whether one understands a phenomenon: if you cannot simulate it, then you do not understand it.[13] Jule Charney wrote

> When a computer simulation successfully synthesizes a number of theoretically-predicted phenomena and is in accord with reality, it validates both itself and the theories—just as the birth of a child who resembles a paternal grandfather legitimizes both itself and its father. (1972, p. 124)

A conspicuous effect of numerical modeling on research style was that parameterization became an important activity. A computer model that represents the atmosphere in terms of the values of certain variables at the nodes of a three- or four-dimensional network (time being one of the dimensions) cannot simulate processes that occur on a scale smaller than the mesh. If the meteorologist wishes nonetheless to include the effects of such a process in his model, then he usually does so by introducing one or more parameters, that is, auxiliary variables that represent statistically the larger-scale or longer-time effects of the process. For example, if the grid points in a horizontal plane are 100 km apart, then thunderstorms, which are typically several kilometers across, cannot be directly represented. Their cumulative effect on, say, temperature or moisture content of the air in a region can, however, be estimated using statistical information about the likelihood, size, and intensity of thunderstorms in a larger region whose average properties are known. Other examples of processes requiring parameterization, if they are to be included in large-scale models of the atmosphere, are the absorption and reflection of radiation, cloud formation, the wave interactions at the air–sea boundary, frictional dissipation of winds, and the gravity waves produced by air flow over obstacles. In the years around 1970 there appeared many journal articles with titles like "A scheme for parameterization of cumulus convection," "Parameterization of surface moisture and evaporation rate," and "Parameterization of the planetary boundary layer."

The example, par excellence, of a process requiring parameterization is turbulence. Since it involves motion on a wide range of spatial and temporal scales, usually the only feasible approach is to represent it as a series of averages, such as mean rate of flow or mean momentum transfer (Drazin and Tveiterind, 1986, pp. 130–131). Turbulence in the atmosphere was remarked on by Helmholtz, as we saw in Chapter 3, and grappled with by Richardson, as we saw in Chapter 6. In recent decades it has been the object of a great deal of study by meteorologists, as well as by applied mathematicians, fluid dynamicists, aeronautical engineers, and others. Its importance in one branch of meteorology was expressed in 1972 by R.W. Stewart as follows: "It is only a little glib, and only marginally stretches the truth, to say that there has been, and remains, only one challenge facing those working with the atmospheric boundary layer: the challenge of parameterizing turbulent phenomena" (p. 270). The effect of computers on the direction of meteorological research can here be seen, since the interest in forecast algorithms and in general circulation models made parameterization a very important mode of research.

While certain methods, such as statistical analysis[14] and bookkeeping studies, seemed to be preadapted to a computer-filled world, other methods, such as the qualitative modeling of the Bergen School, benefited hardly at all from the new computational power. Thus, as computer models became central in meteorological research, synoptic models came to occupy the periphery (Bergeron, 1959, p. 467). Yet the peripheral subjects, although avoided by the meteorologists wishing to devise algorithms, became more attractive to other meteorologists. Jacob Bjerknes in describing his own career wrote that after the demonstration of the effectiveness of computer-forecasting models he "confined his research to the phenomena that could not be very well handled by electronic computer."[15] And the success of computer algorithms for certain types of predictions prompted human forecasters to concentrate on other types of predictions. George Cressman recommended exactly this at a conference in 1967, arguing that the changes in the practice of forecasting due to computers required "a redirection of the activities of professional forecasters into areas least affected by the computer where they can make their most worthwhile contributions." [16]

Computer modeling, as it became the dominant methodology, increasingly gave direction to efforts both in research and in observation: theoreticians often worked to explain the results of simulations (just as they had always worked to explain the results of observation), and observers took the measurements that were needed as input to a model or for testing a model. The work of theoreticians was changed in that an explanation of a simulation was more readily tested than an explanation of an observed phenomenon, and the work of observers was changed in that the measurements had by design a direct relevance to theory. Some reasons for studying computer models rather than the atmosphere directly were stated by Joseph Smagorinsky:

> As the models become more complex, with correspondingly greater degrees of freedom, it has become almost as difficult to determine why the models behaved as they did, as does the real atmosphere. The main advantage in diagnosing model simulations is that we know the physical laws which have been imposed, we know a great deal about the mathematical distortions we have introduced, and right or wrong, we have all of the variables defined everywhere and all of the time. (1970, p. 30)

The influence of numerical modeling on data gathering was commented on by R.W. Stewart in an article surveying work on the atmospheric boundary layer. Stewart wrote, "Many observations will have to be designed specifically to meet the needs of these numerical simulation people," and he stressed the need for "observational programs designed and interpreted specifically with the intent of providing inputs into these models" (1972, p. 278). In this way computer models mediated between theory and data, thus answering to "theory pull." We have already seen how forecast algorithms called for certain types of data and for a regular distribution of observations. In 1954 C.W. Newton published an article on the sensitivity of numerical forecasts to the small changes in the 500-mb chart, and on the basis of his analysis Newton recommended particular improvements in data gathering (p. 287). In 1957 Philip Thompson showed that doubling the density of observing stations would reduce both the initial error and the rate of error-growth (Thompson, 1957). He later wrote [17]

> . . . we had for the first time a quantitative . . . estimate of the increase in predictability to be expected from an increased investment in observing stations, or of the deterioration to be expected from shutting down stations. As it turned out, this was just the kind of ammunition that was needed to save Station Papa, and which was later to provide additional impetus to the planning of GARP and FGGE. (Thompson, 1984, p. 2)

And in 1957 Staff Members of the JNWP Unit reported that certain large errors in forecasts were due to inadequate upper-air data in the Pacific (p. 322).

Numerical Experimentation

Another aspect of the new, computational style of research, in meteorology as elsewhere, was a leading role for numerical experimentation. For centuries meteorological theorizing has been stimulated and curbed by observation and by laboratory experimentation. Observation has always been primary, since the laboratory, although important on occasion, has never been the workplace of many meteorologists.[18] In the mid 20th century, as the electronic computer became available, numerical experimentation grew from a minor, scarcely recognized technique to a standard methodology.

Laboratory scientists traditionally succeeded in gaining understanding of some aspect of nature by constructing and studying in the laboratory an isolated and highly simplified system. The computer made possible a parallel activity in the realm of the symbolic representation of nature: gaining understanding of some aspect of a numerical model by studying the behavior of the model under particular conditions or the behavior of a subsystem of the model. The essence of experimentation—on the computer as well as in the laboratory—is that an external system is put into a chosen state and its subsequent behavior observed. Although numerical experimentation predates the computer—we saw examples in Chapter 3—the advent of the electronic computer made two important differences: the evolution of the numerical system could proceed without human intervention, and vastly more complicated numerical experiments became feasible.

Recall from Chapter 3 Napier Shaw's protocol for meteorological research: find a representation, then study that representation. Shaw had in mind a mathematical representation, a representation that was either symbolic, as Buys-Ballot's law was, or geometric, as Shaw's own model of air motion in a storm was. What the computer did was to make it straightforward to study one class of representations, namely computer models. At the 1955 IAS conference on the study of the general circulation by numerical techniques, Milton Greenberg expressed his belief that by means of the computer one could fulfill "the long-time dream of the meteorologists to be able to study the behavior of the atmosphere under controlled conditions in the same way that experimental models are used in the laboratory" (Pfeffer, 1960, p. 3).[19] And in 1965 Aksel C. Wiin-Nielsen wrote

> . . . the numerical experiment using the non-linear equations has in many instances replaced either the physical simulation of atmospheric and oceanic processes or the greatly simplified "linearized" version which can be treated by classical, analytical methods (in Dingle and Young, 1965, p. iii).

The titles of several talks given at a conference in 1964 suggest the new prominence of numerical experimentation: "Experimental study of micro-structure by high-speed computing technique," "Numerical solution of the distribution of wind and turbulence in the planetary boundary layer," "A mathematical model of air flow in a vegetative canopy," "A numerical study of three-dimensional turbulence using a two-dimensional grid," and "A numerical study of the instantaneous point source diffusion problem."[20]

The new importance of numerical experimentation followed naturally on the practice of constructing numerical models. According to Philip Thompson, "All kinds of questions occur to you that you could answer very easily, that are hypothetical conditions, things that you would normally do in a physics laboratory . . . ," such as increasing the viscosity of the air or the surface friction (1986, p. 32).[21] Indeed, such experimentation was necessary in order to understand computer models. Joseph Smagorinsky wrote

> But as models become more complex, it is difficult, with highly nonlinear and interactive processes, to say why we obtain a given result. There have been many disturbing examples of a result being apparently correct but for the wrong reason. Series of well-designed experiments must be employed to delineate cause and effect. For this purpose, thorough diagnostic techniques must continue to be developed and applied. (1983, p. 37)

George Cressman emphasized the importance of numerical experimentation for developing forecast algorithms: "the prospect of getting a realistic forecast would be completely hopeless without the facility for numerical experiments to isolate and illuminate problems" (1972, p. 182).

Thus, in the first instance, numerical experimentation helps one understand the models. This often leads to changes in the model and a heightened fidelity of model to the observed phenomena, either in the laboratory or in the atmosphere at large. When a model is judged successful, there is reason to think that the numerical experiments on the model give insight into the physical world. Lorenz said of the investigation of hurricanes, "Deeper understanding will come only after further numerical simulation . . . ," (1970a, p. 22), for the reason that one can use a model to answer questions about causation, questions that are difficult or impossible to answer by means of observation.[22]

Another aspect of the new style of research is that it became more common for meteorologists to work as members of a group. Although, as we saw in Chapter 9, forecasting was becoming more and more a team enterprise even before the availability of computers, the development of forecasting methods remained largely an individual enterprise until von Neumann's meteorology project at Princeton. Much of the subsequent development of forecast algorithms—as at the JNWP Unit in Washington and the Institute of Meteorology in Stockholm—was the result of collaboration. Other sorts of research likewise became more collaborative. This is seen clearly in the early work on global circulation models, where the desirability of team research was reflected in the formation of the General Circulation Research Section of the Weather Bureau.

In August 1955 von Neumann, impressed by the results of Norman Phillips's general circulation experiment, drafted a proposal to the Weather Bureau, Air Force, and Navy for a research project on the dynamics of the general circulation.[23] The result was the establishment in October of that year of the General

Circulation Research Section headed by Joseph Smagorinsky. Consisting of five people at the end of 1955, this group grew steadily over the next decades. In 1959 it was renamed the General Circulation Research Laboratory, and in 1963 again renamed, the Geophysical Fluid Dynamics Laboratory. In 1968 it was moved to Princeton. From 1955 to the present it has been a world leader in the new, highly collaborative style of research (Saltzman, 1985, pp. xi–xiii).

This style of research was supported on an even larger scale when representatives of 14 universities established in 1960 the National Center for Atmospheric Research (NCAR) in Boulder, Colorado. The aim was to provide a setting, which the universities separately could not provide, for the collaborative study involving many scientists, often from disciplines outside meteorology, and often requiring expensive computing facilities (Lansford, 1985). With funding mainly from the National Science Foundation and with increased support from universities—there were 54 member institutions in 1985—NCAR soon became the world's largest center for research in the atmospheric sciences.

Numerical Analysis

The use of computers in meteorology, as elsewhere, brought in its train a strong interest in numerical analysis. The rapid growth of this branch of mathematics in the postwar world was largely a result of the development of computers. Its contribution to the success of computational science has often been overlooked. Although the phenomenal increase in the speed of electronic computers in the decades since the ENIAC has attracted much attention, the improvement in numerical techniques—in the ways of reducing sophisticated mathematical operations to the arithmetical operations computers are capable of—has seldom been noted. Yet a study by P.D. Lax at the Courant Institute indicated that in the period from 1945 to 1975 the increase in computing speed achieved by new hardware was less than the increase achieved by advances in numerical methods (cited in Benzi and Franchi, 1988, p. 11).[24] Because numerical analysis was quite a small branch of mathematics before the computer era, scientists using computers often had to invent and improve numerical techniques, which meant that astronomers, physicists, chemists, and other scientists turned into part-time numerical analysts.

In meteorology this diversion toward the mathematics of computation was particularly evident. In 1967 the Russian meteorologist G.I. Marchuk published *Numerical Methods in Weather Prediction.* He wrote: ". . . the principal aims of this book are a discussion of general methods for reducing complicated weather prediction problems to elementary algorithms and a study of the elementary algorithms with regard to their implementation on the computer."[25] In 1972 when two meteorologists, A.J. Robert and M. Kwizak, devised a numerical method of solving the primitive equations that was five times as fast as earlier methods, they remarked, "This is equivalent to acquiring a new computer five times faster" (Kwizak, 1972, p. 1156). In 1978 Philip Thompson commented that "much of the effort in the development of numerical weather prediction over the past twenty years has gone into the improvement of numerical methods," and he gave several examples that show the importance of such work:

> Fairly early in the game it was observed that certain quantities should be exactly conserved, or "invariant," but were not conserved in approximate numerical calculations. This defect was traced to the finite-difference scheme. The U.C.L.A. meteorologist Akio Arakawa subsequently discovered that the latter could be modified in such a way that the known invariants are, in fact, conserved. In 1956, in the course of analyzing the errors of a meteorological calculation, Norman Phillips . . . turned up a new and unsuspected kind of computational instability arising from the nonlinear interactions between wave motions of different scales. (Thompson, 1978b, pp. 148, 150)

Thompson went on to point out that the numerical analysis done by meteorologists benefited scientists in other disciplines, mentioning some important innovations in numerical techniques which had come from meteorologists. He suggested that "meteorology has given back almost as much as it has got from the mainstream of mathematics."

The Unification of Meteorology

As a result of the growth of the science, the availability of computers, and the new styles of research and practice, meteorology became a much more unified discipline. Meteorologists study phenomena over a great range of sizes—from the formation of a fog droplet to tidal movements of the entire atmosphere—and on a great range of time scales—from a lightning discharge to gradual climatic change. These phenomena are of course all related, but it often requires an immense amount of computation to discover the degree and nature of the relationship. Here the computer has been indispensable. For example, studies suggest that the results of the global, long-term simulation may be quite sensitive to the processes of cloud microphysics (Ramanathan, 1988, p. 298). Another example is the mediation between studies of the upper atmosphere and studies of the troposphere that global circulation models provided when, beginning in the mid 1960s, the vertical dimension of some models extended into the stratosphere (Murgatroyd, 1972, p. 149).

The large computer models, either for forecasting or for climate simulations, provided a way of tying together theoretical results in many areas. Successful numerical models of individual processes were often incorporated into a forecasting model or a climate model. Indeed, the research on the individual processes was often motivated by the desire to improve a large model. We saw this with Richardson, and it became common in the 1950s and 1960s. Research in cloud physics was spurred on by Smagorinsky and Collins's 1955 attempt to model precipitation numerically; these researchers had concluded that "Relaxation of the constraints introduced by these approximations depends on progress in the field of cloud physics" (p. 67). In 1967 G.I. Marchuk urged: "Basic investigations in the fields of atmospheric and oceanic physics must be organized primarily to facilitate the application of their results to the problems of weather prediction" (p. 8).

Meteorologists, more than ever before, saw themselves as engaged in a collective enterprise. According to Charney, the promise of the unification of meteorology through computer modeling was "a great psychological stimulus" because all the branches of meteorology "were given new urgency and new importance by the promise that the contributions to the atmospheric circulation from a variety of physical processes could be synthesized mathematically within the computer" (1972, p. 117). In 1970 B.J. Mason wrote

> . . . advances in dynamical meteorology, numerical modeling, and weather prediction have not only greatly strengthened the core of the subject but are beginning to incorporate so much of the rest that the old divisions of synoptic, dynamical and physical meteorology, aerology, and weather forecasting are now becoming largely out-moded and irrelevant. (p. 25)

The wider use of the language of mathematical physics facilitated the unification, and the availability of computers certainly stimulated the devising of physics-based theories. Barry Saltzman wrote

> Not the least effect of the presence of these computers has been the encouragement of conventional mathematical analyses of types which could actually have been performed before the computer's advent (e.g., studies of stability and of first-order energy transfers). (1967, p. 590)

Computational power made physics more useful in the modeling of individual processes, the large-scale motions of the atmosphere (by means of the "primitive equations"), and climatic change. Hence the theories developed by physicists, particularly those studying fluid dynamics, thermodynamics, and atomic and molecular physics, became more important to meteorologists than ever before. It has become more common to regard meteorology as applied physics, and, what is quite new, to so regard climatology as well.[26]

Besides the intramural connections and the connections to physics, computers have mediated connections to other disciplines. There are, of course, the connections, discussed earlier, to computer technology, to software design, and to numerical analysis. The fact that meteorology is only one of many sciences using mathematical models based on fluid dynamics accounts for connections to oceanography, hydrology, aeronautics, astronomy, and other sciences. A mathematical similarity between models in different sciences may raise new questions: Jule Charney, knowing of the work on the solar corona by the astronomers Schatzman and Schwartzschild, asked why the earth does not have an "atmospheric corona" (since a similar mechanism of outward energy propagation might be supposed to operate in the atmosphere) (Charney, 1987, p. 140). Sometimes a similarity between models even provides answers: in 1947 Rossby used the principle of lateral mixing, which he had introduced in 1936 in a study of the Gulf Stream, to explain features of the jet stream, and C.C. Lin's model of spiral nebulae was helpful to meteorologists (Newton, 1959, p. 288; Charney, 1987, p. 132).

As more and more processes came to be included in models of the atmosphere, meteorologists were sometimes able to use the models devised in other disciplines. This has been true of models in hydrology (since hydrologists model precipitation and evapo-transpiration), glaciology, parts of chemistry (especially photochemistry), and parts of biology.[27] Conversely, the existence of detailed atmospheric models made meteorology useful to scientists of many sorts, including astronomers (especially those studying the atmospheres of other planets), historical geologists, and economists (especially for forecasting agricultural production).

The new web of connections is suggested by the great many conferences sponsored jointly by a meteorological organization and an organization from another branch of science or engineering. In the early 1960s the American Meteorological Society shared sponsorship of conferences with, among others, the following organizations: the Division of Fluid Mechanics of the American Physical Society, the American Institute of Physics, the American Mathematical Society, the American Institute of the Biological Sciences, the Instrument Society of America, the American Institute of Aeronautics and Astronautics, the American Society of Civil Engineers, the American Society of Mechanical Engineers, the Society for Experimental Stress Analysis, the Society of Petroleum Engineers, the American Society of Heating, Refrigeration, and Air-conditioning Engineers, and the American Society of Agronomy.

Of all the disciplinary links forged by computational power, that between meteorology and oceanography is the strongest. These sciences have, of course, long been related. From the time of Matthew Fontaine Maury on, there have been numerous institutional ties between meteorology and hydrography, exemplified by the Deutscher Seewarte [German Marine Observatory] in Hamburg and the Nautisk Meteorologiska Buraan [Nautical Meteorological Bureau] in Stockholm, and from William Ferrel on, meteorologists have applied mathematical theories of fluid motion to both hydrosphere and atmosphere.[28] But in recent decades meteorology and oceanography were brought even closer by detailed studies of the air–sea interaction[29] and by the practice of including oceanic motions, not merely a static oceanic boundary, in models of the atmosphere. In 1972 Joseph Smagorinsky wrote

> For climatic time scales it is quite clear that the oceans are just as dependent on the atmosphere as vice-versa. However, the differences in thermal capacity and eddy conductivity result in different relaxation times. It is for this reason that one cannot logically discuss one medium without the other for very long, a realization cogently expounded by Rossby (1957), and observationally demonstrated by Namias (1963) and J. Bjerknes (1966). . . . No numerical model suitable to study the atmospheric general circulation, or to study climate or even to make extended range predictions will be adequate without an active ocean. A simple start in this direction has already been made (Manabe, 1969a, 1969b; Bryan, 1969). (p. 32)

In 1970 B.J. Mason wrote, ". . . before the end of the century, I anticipate that dynamical and physical oceanography will join with meteorology to form a combined discipline . . . ." (p. 33)[30]

In the 1950s and 1960s meteorology became a computational science. This transformation may be said to have started with Vilhelm Bjerknes in 1903, but in the first half of the century computation only slowly expanded its roles in research and practice. Although other factors, especially increasing support for meteorology, were important, the timing and pervasiveness of the transformation were largely determined by the development of computing technology. Computers permitted forecasters to process the data being gathered continuously by satellites and weather stations worldwide, and permitted climatologists to make better use of the vast store of meteorological data by making it easier to search and to analyze statistically. Computers made it easier to apply the laws of physics to the atmosphere and to discover the implications of a set of assumptions; thus physics-based modeling and numerical experimentation became standard methodologies. It was by dint of computation that individual theories were related to data. It was by dint of computation that observation and theory were combined to predict the weather and to simulate climatic change. And it was by dint of computation that meteorology was unified and firmly connected to a host of other disciplines.

Chapter 13 The Recognition of Limits to Weather Prediction

The new computational power that became available in the 1950s and 1960s served in several ways to enhance the image of the science: by making possible marked advances in forecasting and in observational and theoretical meteorology, by facilitating the merger of the three traditions within the science, and by strengthening the connections of meteorology to other disciplines. But the influx of computers into meteorology was not without its casualties. We saw in Chapter 11 how most calculating aids—tables, nomograms, the graphical calculus, slide rules, analog machines—succumbed to the upstart computer. More momentous, however, was the passing away of the belief in the determinacy of atmospheric behavior. Computational power was crucial in bringing about the recognition that what had long been the main goal of meteorologists, to attain predictive success like that of astronomy, was unattainable.

A classic image of a deterministic world was sketched by Pierre Simon Laplace in 1814:

> We ought then to view the present state of the universe as the effect of its prior state and as the cause of the one that will follow. An intelligence which at a given instant knew all the forces by which nature is animated and the respective situation of the things of which nature is composed, and if in addition the intelligence were vast enough to submit these givens to analysis, it would encompass in the same formula the movements of the largest bodies in the universe and those of the lightest atom; nothing for it would be uncertain, and the future, like the past, would be present to its eyes. (p. 2)[1]

But in the 1930s and 1940s as quantum mechanics, according to which certain laws of nature are probabilistic, became an established part of physics, most scientists abandoned the view that the world is fully deterministic, and some saw quantum mechanics as sanction for belief in free will or in a capricious universe. Yet the probabilistic laws seemed to be restricted to the quantum realm, and one could even use them to deduce deterministic laws in the macroscopic realm. Moreover, there was a plethora of physical systems—mechanical, optical, thermodynamic, chemical—that could be shown to evolve deterministically. So most meteorologists, like most scientists generally, continued to view the macroscopic world, certainly the macroscopic physical-world, as Laplace had viewed it and to believe that for meteorology to follow astronomy's example one needed only knowledge of the relevant laws and sufficient computational power. In a 1922 popular presentation of meteorology we read, "A complete knowledge of these laws [that govern the circulation of the atmosphere] would render possible the forecasting of the weather for considerable periods, and thereby meteorology would prove itself to be a science of the greatest value" (Thompson, 1922, p. 763). And in 1959 Tor Bergeron wrote, "Thus, the meteorological forecast stands alone as the most important and promising but still unsolved Laplacian problem on our planet" (p. 441)[2]

Among meteorologists there had long been, it is true, some pessimists who suggested that the weather, more than a day or two in advance, would always be unpredictable. In 1921 the English meteorologist G.M.B. Dobson published a study of errors in the forecasting of winds and pressure gradients. He found that forecasts of winds were generally quite poor and believed that, since small irregular variations in pressure have large effects on winds, there was "little chance of improvement in forecasting in the future" (Dobson, 1921, p. 266).[3] In 1951 Graham Sutton, who 2 years later became director of the Meteorological Office, argued that the weather is inherently unpredictable because small random influences on the atmosphere can have great effects; E.T. Eady made a similar argument in the same year, as did P. Raethjen in 1953, who said that the atmosphere was nearly always in the situation of a "Hercules at the cross-roads" (Bergeron, 1959, p. 440; Smagorinsky, 1970, p. 32).

Although many of the proponents of numerical forecasting believed that computational power would solve the problem of weather prediction, some of the proponents saw computational power as a means of exploring what is and what is not predictable. For example, in 1952 George Platzman wrote

> . . . no matter how thorough or inspired our interpretation of the observations may be, we shall not completely understand the limitations of the classical framework unless we learn a great deal about precisely what can be predicted within this framework; this of course means integration of the (nonlinear) hydrodynamical equations and use of high-speed computers. (p. 178)

The first to investigate the predictability of models was Philip Thompson. In a 1957 paper entitled "Uncertainty of initial state as a factor in the predictability of large scale atmospheric flow patterns," he wrote

> . . . it is an important point of doctrine to know whether or not our uncertainty as to the atmosphere's future state is accounted for by economic (or, ultimately, human) incapacity to observe and compute, or whether it is essential and due to some irreducible minimum of indeterminacy that lies beyond human limitation.

In this paper Thompson showed that "In many winter situations, and for initial error fields whose scale is typical of the present observational network, the inherent RMS [root mean-square] vector wind error may double its initial value after two days, and rise to the error of sheer guessing in about a week."

Thompson's paper attracted only a little attention, and there remained a good deal of optimism about solving the prediction problem. For example, in 1960 Jule Charney expressed the hope that the long-term behavior of the atmosphere might be relatively insensitive to initial conditions, something which is true of many physical systems (p. 15). Then rather suddenly in the mid 1960s the predictability of the atmosphere became a major subject of research. At a conference in 1967 the French meteorologist Guy Dady said, "Finally—we shall stress this point a great deal in what follows—weather prediction has certain limits . . . the most important scientific objective at present being perhaps to specify what is predictable and what is not" (p. 5).[4] And at about the same time, the determination of the predictability of the atmosphere was set as one of the aims of the Global Atmospheric Research Program (Lorenz, 1969, p. 345). Much of the credit for making the question of predictability so prominent must go to the meteorologist Edward Lorenz.

In early 1961 Lorenz was using the Royal McBee LPG-30 computer at MIT to simulate the behavior of the atmosphere over long periods, typically several months. On one occasion he decided to extend an earlier run by starting the computer at a point midway in the earlier run. He expected the first part of the new run to duplicate the last part of the earlier run, but this did not happen—the two runs diverged quite rapidly. Lorenz at first suspected computer malfunction, but then realized that there was a slight difference in the initial conditions of the two sequences. The point midway in the earlier run had been specified, for starting the second run, by the values of variables expressed to three places after the decimal point, whereas the computer kept track of six places (so the earlier run had been continued with the values expressed to six places). It was reasonable to assume that this difference would be inconsequential, since almost all of the physical systems with which scientists were then familiar had the property that when two identical systems were started with very similar initial conditions they would evolve similarly. Lorenz's discovery that with his model of the weather this assumption was wrong led him to a mathematical investigation of deterministic systems that behave in this strange way (Gleick, 1987a, pp. 11–31).

Lorenz soon found a relatively simple mathematical system that had the property of continuing to evolve in a seemingly random way—neither approaching a stable state nor cycling through a sequence of states—and the property of extreme sensitivity to initial conditions. The meteorologist Barry Saltzman had called to Lorenz's attention a system of seven equations describing atmospheric convection that had nonperiodic solutions. Lorenz found he could get the strange behavior with a simplified system consisting of just three equations, and in 1963 published the landmark paper "Deterministic nonperiodic flow."[5]

Lorenz's work linked theoretical meteorology and the mathematical theory of dynamic systems, a subject first studied by Henri Poincaré early in the century (Benzi and Franchi, 1988, p. 8). Moreover, it led to a great change in the way scientists view the physical world. Before Lorenz's 1963 paper most scientists assumed that a physical system would persist in erratic behavior only if it were subject to stochastic law or to erratic perturbations coming from outside the system, and that an isolated deterministic system would eventually settle in at a stable state or assume regular periodic behavior. Lorenz's demonstration of deterministic nonperiodic behavior, which was published in *Journal of the Atmospheric Sciences*, only slowly gained the attention of scientists outside meteorology, but by the mid 1980s his study was being cited in more than a hundred papers each year and the word "chaos," first applied to the behavior of deterministic nonperiodic systems in 1975, had become common in scientific discourse (Gleick, 1987a, pp. 69, 323). Joseph Tribbia and Richard A. Anthes have written, "The ramifications of Lorenz's work have had tremendous impact on the fields of applied and pure mathematics, theoretical physics, turbulence theory, mathematical biology, and the philosophy of determinism" (1987, p. 497). Lorenz's discovery started what some consider to be a major scientific revolution of the 20th century (see Gleick, 1987a, pp. 36–39).

The possible significance for meteorology of the existence of chaotic systems was stated by Lorenz in his 1963 paper:

> When our results concerning the instability of nonperiodic flow are applied to the atmosphere, which is ostensibly nonperiodic, they indicate that prediction of the sufficiently distant future is impossible by any method, unless the present conditions are known exactly. In view of the inevitable inaccuracy and incompleteness of weather observations, precise very-long-range forecasting would seem to be nonexistent.

Lorenz subsequently showed that realistic models of the atmosphere exhibited the same behavior. By the late 1960s the work of Thompson, Lorenz, E.A. Novikov, G.D. Robinson, C.E. Leith, and others had made the predictability of the weather a major subject of research.

The computer was influential in two ways in making the predictability of the weather a subject of scientific inquiry. Obviously it was a valuable tool in the investigation of the predictability of various models. But before one can seek answers to a question, the question must be raised. Here too computers were essential, as Thompson has argued:

> The question of meteorological predictability did not become a real one until the advent of routine numerical weather prediction in 1955. In the public's view, of course, and even in the scientific view, weather had always been regarded as unpredictable, or at least imperfectly predictable, by its very nature. The source of its unpredictability, however, was obscured by the fact that the differences between competing and, at that time, highly subjective predictions were fully as large as the errors of individual predictions. For the first time, with the introduction of numerical methods of prediction, we had an objective standard and could begin to assess the damages of various kinds of error—i.e., errors inherent in the physical models, approximations in the numerical integration of the model equations and errors in specifying the initial state. What I am suggesting is that the question of predictability arose as a natural outgrowth of numerical methods of prediction, and that it wasn't even a very sensible question before they came on the scene. (1984, p. 1)

By 1970 there was abundant support for the view that the atmosphere is not indefinitely predictable. Lorenz and others had shown that meteorologists's inability to make accurate predictions of the weather more than a week or so in advance was not due to ignorance of the relevant laws, nor to limited means of calculation, nor to an inadequate observational network, but to the fact that the atmosphere is a chaotic system, that is, one whose evolution is extremely sensitive to initial conditions.

Here again is a case of meteorological research leading to a great advance in mathematics. Recall the discussion in Chapter 2 of the connection between meteorology and the development of statistics and that in Chapter 11 of advances in numerical analysis stimulated by meteorology. Another example is the invention of partial derivatives, which came out of an effort to deal mathematically with the motions of air: in 1746 the Berlin Academy of Sciences offered a prize for the best paper on the laws governing winds, and d'Alembert's "Réflexions sur la cause générale des vents," which won the prize, introduced the concept of a partial derivative (Khrgian, 1959, p. 217).

We saw in Chapter 6 how Richardson's approach to understanding the physical world led him to mathematical innovation. His work bore additional mathematical fruit decades later when the French mathematician Benoit Mandelbrot chanced to pick up an obscure publication, *General Systems Yearbook*, which contained Richardson's report of a study of how the measured length of a boundary between two countries depended on the scale of measurement. Mandelbrot (1983) generalized Richardson's finding and developed a new kind of geometry—fractal geometry—that treats scale-invariant forms.

A fractal is defined to be a geometrical object with a noninteger Hausdorff dimension. Such objects had turned up several times in mathematical research, but they were regarded as isolated "pathological" types having no possible application to the physical world. Mandelbrot's theory has shown both that these objects are related to one another and that they have significant applications in several sciences. Fractal geometry has indeed become a standard tool in physics and is perfectly suited to treat another type of scale invariance Richardson had called attention to—turbulent motion in the atmosphere.

It turns out that there is a connection between chaos and fractals: the trajectory in state space that describes the evolution of a dynamical system delimits a mathematical form called an "attractor," and the dynamical system is chaotic exactly when the attractor is a fractal (in this case called a "strange attractor"). Richardson's work is thus connected to chaos by two channels of development: mathematical modeling of the atmosphere (culminating in Lorenz's work) and mathematical modeling of scale-invariant forms (culminating in Mandelbrot's work).[6]

The point is that a persistent effort to discover a mathematical way to describe and predict refractory phenomena may lead to mathematical innovation and that meteorology has been a great provider of refractory phenomena. We have seen in earlier chapters that the mathematical innovations have in turn served meteorology well, and the discovery of chaos is no exception.

The effect on meteorology of the discovery of chaos may be summarized as follows. There were two traditional programs to achieve long-range prediction of the weather. The program that was almost always dominant may be called the nomothetic program, since scientific law was its mainspring. According to this program three things were needed: (1) the ability to specify the present state of the atmosphere, (2) knowledge of the laws governing atmospheric action, and (3) the calculational means of applying the laws to a description of the present state in order to deduce future states of the atmosphere. Empiricists, such as Julius Hann and Napier Shaw, thought that (1) and (2) presented the main difficulties and that progress with (1), together with an accumulation of the descriptions of the atmosphere, would lead to progress with (2). Bjerknes, Richardson, and von Neumann saw (3) as the main difficulty.

The other program to achieve long-range prediction may be called the historical program, since its basis was an archive of descriptions of past states of the atmosphere. This program too had three requisites: (1) the ability to specify the present state of the atmosphere, (2) a vast archive of descriptions of past states of the atmosphere, and (3) the ability to select from the archive the past state most closely resembling the present state. Meteorologists enlisted in this program usually emphasized the need to build up the archive, although they continually exercised ingenuity in devising ways—involving such aids as indexes, punched-card machines, and analog computers—to make the selection from the archive.

Meteorologists have always recognized certain limitations on the ability to specify the state of the atmosphere at a particular time: the instruments of observation have limited accuracy, and, moreover, since they cannot take measurements everywhere simultaneously, some sort of interpolation must be made. The discovery of chaos and the demonstration that the atmosphere is probably chaotic called attention to these unavoidable limitations on description of the atmosphere, and they were seen as severely constraining both the nomothetic and the historical programs. The nomothetic program now has less ambitious aims because for realistic models of the atmosphere one can show, both numerically (with a computer) and analytically, that the growth rate of error is roughly exponential. This means that the smallest errors will rapidly become large, even if the model and the computation are exact. Similarly, the historical program looks less promising: it seems unlikely that the atmosphere would ever have exactly the same description twice, and one can show that even miniscule differences—differences too small to be measured by any conceivable observational devices—in the initial state lead quickly to large differences in subsequent states.

Although it seems that neither the nomothetic nor the historical program can yield detailed and complete predictions that are accurate for more than a week or two, there is the possibility that certain aspects of the weather—perhaps the average temperature or the total precipitation over an entire season—are predictable months in advance, and adherents to both programs hope to demonstrate this predictability.[7]

The use of the computer has led to a much better understanding of what is and what is not predictable, of degree of predictability, and of how predictability depends on the scale of the phenomenon. More generally, numerical studies have permitted estimation of the value, for predictive accuracy, of possible improvements in observation, theory, or computational power. Thus the computer, besides being a powerful engine in moving observational, theoretical, and practical meteorology along, has also served as a navigational aid to help meteorologists and those funding meteorology decide where to steer.

Notes

Chapter 1

1 In 1932 the tripartition of the science was described by Napier Shaw, the doyen of English meteorologists, as follows: "And by meteorology I would have the reader understand not merely the statistical results of the accumulated observations of weather which may be incorporated with physical geography, nor the process of prediction of the weather of to-morrow, but progress towards the comprehension of the physical and dynamical principles of the structure and general circulation of the atmosphere and of the changes that are liable to occur therein" (Shaw, 1932, p. 394). See also the epigraph to Part I.

2 This distinction is not, however, carried over to the agent nouns "computer" and "calculator." A different distinction is made here: a calculator is a device for doing single mathematical operations, whereas a computer is a programmable device. (The designation "programmable calculator" has arisen because "calculator" is firmly associated with the pocket-sized device, which originally did perform only single operations.)

Chapter 2

1 Not even on 4 July 1776 did Jefferson neglect his weather journal: on that day he recorded the temperature at 6 a.m. (68°F), 9 a.m. (72¼°), 1 p.m. (76°), and 9 p.m. (73½°) (Moore, 1898, p. 7).

2 It is interesting that for the Greeks both astronomy and meteorology were sciences of direction: astronomy dealt almost exclusively with the direction of heavenly bodies, and most meteorological observations were of wind direction, presumably because of its importance in navigation. However, rather than quantifying wind direction, the Greeks indicated it by assigning one of some 10 names to a wind.

3 The most complete account of this transformation is W. E. K. Middleton's *Invention of the Meteorological Instruments* (1969). Other accounts are Theodore Feldman's *The History of Meteorology, 1750–1800* (1983), and A. Kh. Khrgian's *Meteorology: A Historical Survey* (1959).

4 It is interesting that meteorologists succeeded in quantifying wind speed long before they succeeded in measuring it accurately. In the first decade of the 19th century Francis Beaufort devised a wind scale according to which a number (from 0 to 12) was assigned to the wind according to qualitative observations (such as whether wind is felt on the face or whether crested wavelets form on water). When meteorologists did develop accurate anemometers, the wind speed v corresponding to a Beaufort number B could be measured; Henrik Mohn in the February 1890 *Meteorologische Zeitschrift* reported such results. The relationship is approximately $v^2 = k\,B^3$, where k is constant (Shaw, 1926, p. 18).

5 It was especially important to establish conventions concerning the exact placement, the "exposure," of instruments. For example, the reading of a thermometer is particularly sensitive to such factors as the amount of ventilation, proximity to the ground or buildings, and exposure to direct or reflected sunlight.

6 The organizational transformation of the science is the subject of James Fleming's *Meteorology in America, 1800–1870* (1990). In the 19th century meteorologists shared with many other scientists the concern for communality of data. For example, the constitution of the American Metrological Society, organized in 1873, states, "The primary object of this Association shall be to improve the systems of Weights, Measures and Moneys, at present existing among men, and to bring the same, as far as practicable, into relations of simple commensurability with each other" [*Bulletin of the American Metrological Society,* Vol. 5 (1879), p. 5].

7 Form 1009 continued to be used, although quite a few changes were made in 1949 (U.S. Weather Bureau, 1963, p. 4). In 1985 Jack Thompson wrote that the daily observations made when he began work for the Sacramento (California) office of the Weather Bureau in the late 1920s "did not differ materially from the present synoptic observation" (p. 1251).

8 Something of an outline of the history of numerical tables is contained in R. C. Archibald's *Mathematical Table Makers* (1948).

9 A new collection, partly based on Guyot's, was published by the Smithsonian Institution in 1893, and later editions of it appeared in 1896, 1897, 1907, 1918, and 1939.

10 In Conrad and Pollak's *Methods in Climatology* (1950), for example, "Climatological examples demonstrate the computations so that everyone with high-school training should be able to understand the mathematical procedure" (p. xi).

11 Some of the things people expected to find associated with particular types of weather were astronomical events, sickness, harvests, and economic activity.

12 The single exception was "that the barometer dipped briefly for a small fraction of a line as the sun crossed the meridian at noon and midnight" (Cassidy, 1985, p. 22).

13 Parts of the history of pictorial and graphical techniques are recounted in Robinson (1982), Funkhouser (1937), Beniger and Robyn (1978), and Tufte (1983).

14 The earliest citation in the Oxford English Dictionary of "graph" in the sense of the graphical representation of the locus of a function is 1885.

15 These figures, numbers 2, 3, 9, and 10 in Hann (1883), are more akin to a geologist's or geographer's vertical section than to the graphs of data common today.

16 Although seldom thought of in this way, drawing isolines is itself a form of graphical calculation; it is interpolation in two dimensions.

17 ("Unter Klima verstehen wir die Gesamtheit der meteorologischen Erscheinungen, welche den mittleren Zustand der Atmosphäre an irgend einer Stelle der Erdoberfläche charackterisieren.")

18 ("Eine wissenschaftliche Klimatologie muss danach streben, alle klimatischen Elemente durch Zahlenwerte zum Ausdruck bringen zu können, da nur durch wirkliche Messung unmittelbar vergleichbare Ausdrücke und bestimmte Vorstellungen der meteorologischen Verhältnisse und Zustände gewonnen werden können.")

19 ("Aber bei aller Veränderlichkeit hat das Wetter in jedem Orte der Erdoberfläche einen bestimmten allgemeinen Charackter, der sich nicht nur in Durchschnittswerten und Extremwerten, sondern auch in der Art des Wechsels, im ganzen periodischen und unperiodischen Verlauf des Wetters ausspricht. Diese Gesamtheit der Witterungserscheinungen eines Ortes nennen wir sein Klima.")

20 According to Porter (1986, p. 273), "Galton's ideas about the correlation of variables probably derived in considerable measure from his meteorological work."

21 Although the word "climate" goes back to ancient Greece, "climatology" was not coined until the early 19th century. "Klimatologie" appears in the title of a book, written in German by the Danish scientist F. J. Schuow, that was published in Copenhagen in 1827. One of the earliest uses of the word in French occurs in a book by Humboldt published in Paris in 1841. (This predates by 2 years the earliest citation in the *Grand Larousse de la langue française*.) The earliest citation in the Oxford English Dictionary of "climatology," "climatography," or other forms of these words is from 1843.

22 From pages 144 and 178 of Ward's 1903 translation of the first volume of the second edition of Hann's *Handbuch der Klimatologie*.

23 Examples are the following: "Rapid descent warms air while rapid ascent cools it," "Winds are deflected by the rotation of the earth; to the right in the northern hemisphere and to the left in the southern," and "The rate of evaporation increases with temperature and wind velocity, and with decreased relative humidity."

24 Biology was directly tied to climatology in phenology, the study of the relations between weather and annual biological phenomena such as plant flowering or bird migration.

Chapter 3

1 The same derivation was made independently, and only slightly later, by C. M. Guldberg and Henrik Mohn (see Kutzbach, 1979, p. 101).

2 According to the English meteorologist Ernest Gold (1930, p. 195), "Before the publication of this book [Exner's *Dynamische Meteorologie*] the student of meteorology who was interested in the dynamical aspect of the subject had to refer to original papers scattered through many journals in different languages before he could get a grasp of the work which had already been done in that field." He goes on to say that the only exception to this was the collection of papers translated by Cleveland Abbe. Three such collections were published by the Smithsonian Institution, in 1878, 1891, and 1910. In 1885 Adolf Sprung published *Lehrbuch der Meteorologie,* which has been called "the first complete work on dynamic meteorology" (Grunow, 1975, p. 595), but this was at an early stage in the development of the subject. The widely read *Lehrbuch der Meteorologie* (1901) by Julius von Hann is essentially descriptive and leaves out "the investigations in mathematical physics which form the theoretical foundation of the subject" (Walker, 1925b, p. 430). H. H. Hildebrandsson and L. Teisserenc de Bort's two-volume *Les bases de la météorologie dynamique* (1900, 1907) is, despite the title, descriptive.

3 *Proceedings of the American Association for the Advancement of Science*, Vol. 39, p. 77 (1890).

4 From 1900 to 1905 Shaw's title was Secretary of the Meteorological Council; the Council was the governing board of the Meteorological Office. In 1905 the position of Director of the Meteorological Office was created, and Shaw held that position from then until his retirement in 1920.

5 Similar comments about Shaw's effect on the Meteorological Office are made in Sutton (1955).

6 The Oxford English Dictionary cites no earlier use of "model" in the sense of a "simplified or idealized description . . . of a particular system, situation, or process . . . that is put forward as a basis for calculations, predictions, or further investigation" than Niels Bohr's 1913 use of the word in referring to hypotheses concerning atomic structure. There is an important distinction between explaining some phenomenon directly (imagining that one's words are directly applicable to reality) and proposing a model for the phenomenon (presenting a mental or symbolic construct). In the former case language tends to be invisible: arguments are thought to be about the phenomena. In the latter case the mental construct (that is, the model) is put forward as an object. There are then two quite different activities: study of the model in the logical realm and study of the corre-

spondence between the model and the phenomenon in question. And criticisms may stay within the logical realm (arguing, for example, that the model is internally inconsistent or inconsistent with an accepted law) or may focus on the manner and degree of correspondence between model and reality. It is only in proposing a model that one is conscious of constructing something. One can then manipulate this construction to achieve a better correspondence between what the model predicts and what in reality happens. This was clearly what Shaw, Margules, Richardson, and others were sometimes doing, and doing consciously.

7 Shaw was writing this several years after the celebrated confirmation of the general theory of relativity by solar eclipse measurements. Einstein's revision of Newtonian mechanics Shaw took as confirming his views that (1) a person should not assume that the known laws of physics suffice and (2) a person should work from the data. Einstein would no doubt have agreed with (1), although hardly with (2).

8 Half a century earlier Humboldt had said that summaries of numerical data were the great inheritance that one century gave the next (Sheynin, 1984, p. 66).

9 See, for example, the foreword to Alfred Wegener's *Thermodynamik der Atmosphäre* (1911), or the introduction to Exner's *Dynamische Meteorologie* (1917), or the preface to Brunt's *Physical and Dynamical Meteorology* (1934). The phrase "contact with reality" occurs in a review of Exner's book (Walker, 1925b, p. 431).

10 Four years later Margules used a series of calculations to establish the plausibility of an alternative theory, that the kinetic energy of storms came from the potential energy of the juxtaposition of warm- and cold-air masses.

11 Helmholtz's paper is translated in Abbe (1891).

12 Ferrel's equation is $G = s\,P\,k\,[2n \cos \psi + (s \cos \phi \,/\, r)] \,/\, [\cos \phi\,(1 + mt)\,P']$, where G is the barometric gradient, s is the speed of the wind, P is the atmospheric pressure, P' is the pressure of a 760-mm column of mercury, k, m, and $2n$ are numerical constants, ψ is the complement of the latitude, ϕ is the inclination of the wind's direction to the isobars, r is the distance from the center of the low-pressure area, and t is the temperature (Loomis, 1883).

13 ("C'est ici surtout que se fait sentir la nécessité d'employer un très-grand nombre d'observations, de les combiner de la manière la plus avantageuse et d'avoir une méthode pour déterminer la probabilité que l'erreur des résultats obtenus est renfermée dans d'étroites limites, méthode sans laquelle on est exposé à présenter comme lois de la nature les effets des causes irrégulières, ce qui est arrivé souvent en Météorologie.")

Chapter 4

1 The example comes from the cuneiform library of the Assyrian king Ashurbanipal (668–627 B.C.) (Hardy, 1982, p. 182). One of the earliest compilations of weather signs is that of Theophrastos, written about 310 B.C. David Brunt wrote that "many of the Weather Signs of Theophrastus are to be found in the weather lore of widely separated countries, in almost identical words" (1951, p. 114). Sources of information about traditional weather signs include R. L. Inwards's *Weather-lore* (1869/1950), H. H. C. Dunwoody's *Weather Proverbs* (1883), S. K. Heninger's *A Handbook of Renaissance Meteorology* (1960), and Charles Galtier's *Météorologie populaire dans la France ancienne* (1984).

2 There were a few meteorologists who constructed such maps before the telegraph made possible same-day maps. As early as 1816 the German physicist H. W. Brandes argued that one could learn much from a series of maps of the daily weather, and he used data from 1783 to construct such maps (Schneider-Carius, 1955, pp. 156–161). In the 1840s in the United States, James P. Espy and Elias Loomis published synoptic charts.

3 At the turn of the century, the U.S. Weather Bureau received daily reports from about 200 stations (Moore, 1898, p. 12) and the Austrian weather service from more than 140 stations (Pernter, 1903, p. 161).

4 Examples of such charts for Western Europe are in Köppen (1882) and van Bebber (1891).

5 From 1855 until 1867 the British weather service was the Meteorological Department of the Board of Trade; in 1867 its name was changed to Meteorological Office (and oversight given to a committee nominated by the Royal Society) (Brunt, 1951, p. 117).

6 The committee reported, "We can find no evidence that any competent meteorologist believes the science to be at present in such a state as to enable an observer to indicate day by day the weather to be experienced for the next 48 hours throughout a wide margin of the earth's surface" (in Hughes, 1988, pp. 203–204).

7 Mallock ventured the opinion that it was "extremely unlikely that any trustworthy weather forecasts for periods so long as twenty-four hours will, or can, ever be made for regions outside latitudes 30° N. or S. . . ."

8 It was especially in the construction of auxiliary charts that calculation was done. The isallobaric chart, first proposed by N. Ekholm in 1906, showed the geographic distribution of rate of change in barometric pressure (Neis, 1956, p. 36). Many forecasters prepared temperature-change charts and humidity charts (U.S. Weather Bureau, 1916, p. 77; Moore, 1898, p. 12).

9 This method assumes the precept that if two identical systems start in two very similar states then they will evolve similarly. As discussed in Chapter 13, in the 1960s scientists discovered that for certain systems, called "chaotic," this precept is false, and meteorologists found reason to believe the atmosphere chaotic.

10 Several articles on these efforts, including W. V. Brown's "A proposed classification and index of weather maps as an aid in weather forecasting," appeared in the December 1901 issue of *Monthly Weather Review*.

Chapter 5

1 It appears, however, that Exner changed his mind on this point, for in the introduction to his *Dynamische Meteorologie* (1917) he writes that meteorology is an explanatory science, but not a predictive science because not all the factors that determine the weather can be treated theoretically.

2 The seven equations, in the form Richardson used them in *Weather Prediction by Numerical Process* (1922), are given in Figure 4 of Chapter 6.

3 This despite the fact that Bjerknes made much use of numerical tables. Some 80 pages of Bjerknes and Sandstrom's *Statics* (1910), for example, are given to tables for computing.

4 Every student of calculus has noticed that graphical solutions are frequently easier than analytic solutions. Drawing a tangent is usually easier than differentiating; counting squares is usually easier than integrating. Equipped with a single diagram on transparent paper of nested circles all tangent at a point, one can quickly determine graphically the radius of curvature of a function; to find radius of curvature analytically one evaluates $[1 + (dy/dx)^2]^{3/2} / (d^2y/dx^2)$. Figure 4 of Chapter 6 shows one context in which meteorologists must determine radius of curvature.

Chapter 6

1 "Letter" is defined to be a publication of four pages or fewer. (The underlined entries involved some guesswork about whether a publication should be classified as a letter or as an article.) Most

of the information in this table on employment is taken from Gold (1954), and most of the information on publications is taken from Appendix A of Ashford (1985).

2 Some other computers recruited from the family were Blaise Pascal and John Mauchly, who as children did calculations for their fathers (and who invented computing devices), and the wives of the meteorologists Jule Charney and Arnt Eliassen. In the computer era, it was often programming rather than calculating that family members contributed: Klary von Neumann, Kathleen Mauchly, and Margaret Smagorinsky wrote machine code for their husbands.

3 Eskdalemuir Observatory was built by the National Physical Laboratory and in 1910 transferred from the NPL to the Meteorological Office.

4 It was not so much scientific curiosity as a sense of duty that motivated Richardson here. Near the end of his life Richardson wrote, "A persistent influence in my life has been that of the Society of Friends [Quakers] with its solemn emphasis on public and private duty" (in Ashford, 1985, p. 1).

5 Anatol Rapoport calls Richardson's view—that wars are brought about by diffuse social forces—Tolstoyan. This view he sets in opposition to the Clausewitzian—that wars result from rational foreign-policy decisions (Rapoport, 1968). The latter view is modeled by game theory rather than by differential equations.

6 Richardson might have avoided the errors caused by the intermixture of plus and minus signs if he had organized his computers as was done several decades later in a Works Progress Administration project in New York City (Ritchie, 1986, p. 15). There the (human) computers were divided into four groups, the first group did addition only, the second subtraction only, the third multiplication only, and the fourth division only. Positive numbers were written in black, negative numbers in red. The instructions given to the first group were "Black plus black makes black. Red plus red makes red. Black plus red or red plus black, hand the sheets to group 2."

7 The asterisk is a reference mark for the footnote "Carnegie Institution, Washington, 1910, 1911."

8 Bjerknes *et al.* had earlier presented data at various stations and for various heights in a similar tabular arrangement (1911, p. 20).

9 Indeed, it was to faulty data on winds high in the atmosphere that Richardson attributed much of the error in the forecast he computed (see also Richardson, 1924a).

10 The observer, it is perhaps needless to say, is protected by a roof through which the barrel of the gun projects. Nevertheless, the danger associated with this observational technique is probably the reason why it was never much used. [Richardson's instructions on doing these observations did include the following: ". . . it is important to call out before firing 'Hullo! Danger! Bullets will fall! Keep off!' or something to that effect" (in Ashford, 1985, p. 129).]

11 When Richardson designed an observational instrument to measure a particular variable, he worked in two ways to eliminate the physical effects that interfered with the measurement: he designed the instrument so that these physical effects did not occur, or he made a separate measurement of each effect and then subtracted it from the raw data. The first way is exemplified by Richardson's care in constructing a sensitive electrical thermometer that could be raised to different heights above the deck of a ship (Richardson, 1927, p. 296); he designed the thermometer so that it was not subject to heating or cooling by rain, radiation, friction, or adiabatic compression. The second way is exemplified by the elaborate calculations in Richardson's paper on measuring wind speed by shooting spheres upward.

12 The motions at the molecular level are taken into account by the quantity assigned to a fluid that is called its viscosity. (Thus there is not an infinite regress.) The macroscopic effects of these motions depend on the temperature, pressure, and little else, so that viscosity—unlike turbulence—can be considered a property of the fluid.

13 Frank Gifford (1972) points out that, although unknown to Richardson, energy is fed upscale in part of the range, from synoptic storms to the general circulation. He accordingly offered the following second stanza: "And the great whirls in turn supply/ still greater whirls' rotation;/ And these feed greater still, up to/ the general circulation."

14 A dimensionless quantity is one in which, when one computes it, all physical units (centimeters, seconds, degrees Kelvin, amperes, and so on) cancel out. This means that the number comes out the same regardless of the system of units used in computing it.

15 In 1925 Exner published the second edition of his *Dynamische Meteorologie*. In this book, commenting on a line of work being pursued in England on the details of air motions in turbulence, Exner writes, "The work of L. F. Richardson, Weather Prediction by Numerical Process, represents this direction to a large extent. I have not succeeded in incorporating this valuable but difficult work in the new edition: the general formulation of the problem is too different" (in Platzman, 1967, p. 541).

16 On page 146, for example, Richardson (1922) writes: "But it looks as though equation (18) would be found to be inaccurate. As I have carefully checked its deduction, I suspect the error to reside in the hypotheses."

17 In general, the equations give the time rate of change of a variable in terms of spatial rates of change. The latter are known from the spatial distribution of initial data. (Actually, of course, the spatial derivatives are replaced by ratios of spatial finite-differences.)

18 For example, on page 42 he writes: "To fit with the rest of this work we require the entropy expressed as a function of pressure, density and moisture; the temperature must not appear explicitly. However, there is a difficulty in expressing the relation in formulae, because the vapour pressure is given experimentally as a somewhat complicated function of the temperature. On this account it is simpler to proceed by way of graphs or numerical tables, of which those by Neuhoff appear to be the best, although they need to be converted to millibar units."

19 This quotation suggests the recognition of a type: the English scientist who was brought up on the classics. Some characteristics of this type of humanity are the use of Greek words (written in the Greek alphabet), Greek and Latin epigrams (untranslated), Latin titles (Napier Shaw wrote *Principia atmospherica*), classical allusions, and learned coinages (if Shaw had had his way, we would now be saying "anaphalanx" and "kataphalanx" instead of "warm front" and "cold front"). Other characteristics are a precision in the use of Latinate English words, a tolerance that is sometimes too clearly an expression of a magnanimous nature, a predilection for a historical presentation of the science, an elevated tone, a universal interest, and a complete ease in making lengthy digressions. (Richardson gently chides Shaw for these last three characteristics in the following passage from a book review: "Portions of information which seem to have come from one chapter have diffused into other chapters, like cumuli into the blue sky. Sir Napier "bloweth where he listeth," and it would be hard to tell whence he cometh or whither he goeth, were it not that there is a full index and a summary of the contents of the four volumes.") Augustus DeMorgan and Archibald Geikie are other examples of this type of humanity.

20 The American equivalent is a Rube Goldberg device. The early decades of this century were a time when people were getting used to being surrounded by machines. One might learn a lot about popular attitudes toward machines, and popular understanding of mechanisms, by studying the works of Heath Robinson, Rube Goldberg, and Storm Petersen (who did similar work in Denmark).

21 Herman H. Goldstine has written *A History of Numerical Analysis from the 16th Century through the 19th Century* (1977), which is meticulously documented and mathematically complete. Goldstine was himself a central figure in the development of electronic computers and one of the major contributors to numerical analysis (when it was a fledgling field and later), and he has written the important book *The Computer from Pascal to von Neumann* (1972). But the story of the relationship in the 20th century between numerical analysis and computers remains to be written.

22 On page 54, for example, is the following: "It would be interesting to repeat the computation of the absorptivity on the assumption that dry air is perfectly transparent except between wave lengths of 13 and 16 microns, which is the position of the band due to carbon dioxide."

23 Richardson later wrote (1932, p. 220): "Einstein has somewhere remarked that he was guided towards the theory of relativity partly by the notion that the universe is essentially simple. R. H.

Fowler has been heard to say that of two theories the more elegant is probably the more physically correct. If such eminent physicists could be persuaded to attend to meteorology, that science would be greatly enriched; but they would probably be forced to abandon the faith that Nature is essentially simple."

24 These theories are discussed in Richardson (1922) on pages 23, 27, 30, 30, 50, 61, 104, 112, 127, 134, and 140, respectively.

25 These measurements are presented in Richardson (1922) on pages 23, 32, 41, 41, 44, 72, 86, 121, 134, and 145, respectively.

26 Astronomers, to be sure, carried out computations that were much longer, and mathematically more complex, but most of these computations involved mainly a single phenomenon—gravity—and a single force-law.

27 On page 128, for example, Richardson (1922) arrives at certain equations describing the winds in the stratosphere. He writes: "To test this theory I selected the two highest balloon flights included in V. Bjerknes' Synoptic Charts, hefts 1,2,3." Another example is the work reported in Richardson *et al.* (1922b), which was a test of a theory used in *Weather Prediction by Numerical Process*.

28 Because the schemes of numerical weather prediction used since 1950 closely resemble Richardson's (in particular, they all partition the atmosphere in the same way), this reasoning still obtains. For example, in a 1978 article Philip Thompson (1978b, p. 148) wrote that "decreasing the linear dimensions between grid points by a factor of two results in a sixteen-fold increase in the total number of numerical and logical operations," which is another way of expressing the n^4 term.

29 Elsewhere he estimated the value of the world's food production to be at least a billion pounds sterling, "so that a tiny fractional saving would correspond to a large sum" (in Gold, 1954, p. 223).

30 This despite what he said about the "houi" of the machine prompting him to continue. The sheer labor of calculation must have dissuaded him. In a letter to Shaw, Richardson asked about an idea of his: "Is this any good? I see no prospect of ever finding time to work it out in detail. And am much more attracted by a prospective Swiss holiday than by calculations" (in Ashford, 1985, p. 141).

31 Richardson the *meteorologist* was not an ignored genius. With his peace research the story is different. Here he had almost no audience at all. If *this* work had been widely noticed, much might have been made of it at the time. Much was made of it in the 1950s and later.

Chapter 7

1 David Ludlum discusses at length the effect of weather on fighting in the War of American Independence, the War of 1812, and the War Between the States in his book *The Weather Factor* (1984).

2 It is true that in the United States the national weather service was provided by the Signal Corps of the Army from 1870 until 1891. But in 1886 a joint House–Senate committee concluded, "As the scope of the weather service has enlarged, the military feature has become unimportant. . . . The Army gets no benefit from this Signal Corps, and places no reliance upon it for any military service" (in Bates, 1956, p. 521). In 1891 the weather service was transferred to the Agriculture Department.

3 Military aviation during World War I did much to stimulate the development of meteorology (see Friedman, 1978). During the war the military established many meteorological stations especially for aviation. [Another new type of station was the "winds-aloft station" for artillery. The physicist Alan T. Waterman (Director of the National Science Foundation from 1951 until 1963) worked as a meteorologist for the U.S. Army in World War I and helped to set up both these types of weather

stations, an aviation station at Langley Field (in Virginia) and a winds-aloft station at Fort Monroe (also in Virginia) for the Coast Artillery (Waterman, 1952, p. 186).]

4 During the war the British Admiralty, War Office, and Air Ministry each established a meteorological service. In 1920 these services were incorporated into the Meteorological Office, which itself became, in a sense, part of the military as it was transferred to the Air Ministry. One result (as mentioned in Chapter 6) was that Richardson, a conscientious objector during the war, felt compelled to resign his position with the Meteorological Office.

5 A higher-level rule, although not connected to physics at the time it is formulated, may become so connected, as was the case with Buys-Ballot's law, discussed in Chapter 2.

6 An air mass is defined to be a region of the atmosphere, extending typically 1000 km horizontally and a few kilometers vertically, having roughly constant physical properties as a result of having been situated for some time over an ocean or generally flat continental area. An occluded front is one cut off from contact with the surface of the earth; it is usually a warm front forced aloft by colder air.

7 The slowness of the French and Germans to adopt the Bergen methods is mentioned in Reichelderfer (1970, p. 208). The missionary activities of Tor Bergeron on behalf of the Bergen School are recounted in Eliassen (1978); Bergeron was particularly successful in Russia.

8 The person who was most influential in the decision by the U.S. Weather Bureau to adopt these techniques was Carl-Gustaf Rossby, and yet it was Rossby who, in the succeeding two decades, did most to reintroduce physics into forecasting. (Vilhelm Bjerknes moved from a physics-based meteorology to one independent of physics; Rossby countermarched.)

9 Kelvin's statement that "when you can measure what you are speaking about, and express it in numbers, you know something about it; but when you cannot measure it, when you cannot express it in numbers, your knowledge is of a meagre and unsatisfactory kind" has been a favorite quotation among dynamical meteorologists. It is, for example, the epigraph of Petterssen's *Weather Analysis and Forecasting* (1940). It was quoted in the conclusion of a memorial lecture on Jule Charney delivered by Norman Phillips in 1982 (p. 497), and in a 1953 article it is said to be too famous to quote (Gringorten, 1953, p. 59). Merton, Sills, and Stigler (1984) have written a short history of the Kelvin dictum, which long predates Kelvin, in the social sciences.

10 In 1907 Arthur Schuster had established a Readership in Meteorology at Cambridge (Brunt, 1951, p. 123).

11 This report mentions that Oliver L. Fassig, instructor at Johns Hopkins University, was promoted to "associate in meteorology" and that Robert De Courcy Ward was "assistant professor of climatology" at Harvard University.

12 In the 1890s Johns Hopkins University had a short-lived graduate program in meteorology; one Ph.D. was awarded, to Oliver L. Fassig in 1899 (Koelsch, 1981). In 1923, and perhaps earlier, Harvard University had an undergraduate "concentration" in meteorology and climatology [*QJRMS* **49,** 136 (1923)].

13 Whitnah wrote (1961, p. 167), "Aid to American aviation has involved more innovation, technological development, and opportunities for expansion of Weather Bureau facilities than any other field of weather service." An important effect of forecasting for aviation is summarized by Friedman (1978, p. 227): "In 1914, when the war broke out, meteorologists were drawing and studying weather maps based on observations that distinguished between merely five types of weather and no observations of cloud types. By 1922 one hundred possible varieties of weather, over fifteen cloud forms, and visibility were routinely observed and made available for analysis on weather maps. This drastic change, which proved essential for creating indirect aerology and subsequently the air-mass concept, resulted from aviation's meteorological needs." The more detailed and constantly revised forecasts required for aviators came to be appreciated by other users of forecasts, such as farmers, shippers, and promoters of outdoor entertainment [*BAMS* **18,** 40 (1937)].

14 A radiosonde is a miniature radio-transmitter connected to sensors (such as thermometers, barometers, or gas sensors) which, when carried aloft by balloon or rocket or dropped from an airplane, broadcasts signals encoding the readings of the sensors.

15 About 20% of the initial membership of 600 were in academia, 40% were employed by the government, most of them in the Weather Bureau, and most of the remaining 40% were amateur meteorologists (Reichelderfer, 1970, pp. 207, 210).

16 This book, published in 1938 by the Institute for Research in Chicago, recommends that one choosing to become a meteorologist be a "particularly stable and unsensitive type" since meteorologists and their forecasts are continually joked about. It recommends also that the person not have a distaste for routine work, which is an indication of the amount of algorithmic data-processing in meteorology at the time.

17 ("Il est certain, en effet, que les météorologistes font une séparation radicale entre la théorie et la prévision.") Giao was proposing a new principle, a radical version of the Bjerknes program: "The value of a theory of dynamical meteorology is measured only by its applicability to the quantitative forecasting of weather." ("La valeur d'une théorie de météorologie dynamique ne se mesure qu'à son applicabilité à la prévision quantitative du temps.")

Chapter 8

1 Another WPA project, the New York Mathematical Tables Project, which employed about 350 people to carry out computations, did some work for meteorologists.

2 A duplicating punch is a manually operated key punch with a duplicating facility, so that data common to several cards can be punched automatically while the data specific to the individual card are punched manually. An interpreter converts punched-card information into printed information.

3 This fascination is today manifested in an organization called the Foundation for the Study of Cycles, which publishes two journals, *Cycles* and *Journal of Interdisciplinary Cycle Research.*

4 In recent decades the search for cycles of very long period, thousands of years or more, has become an important scientific activity, one conducted as much by geologists as by meteorologists. Cycles of such long period are denominated climatic cycles, rather than weather cycles.

5 W. J. Humphreys (1942, p. 367) saw this as "the beginning of the innumerable substantially fruitless studies of weather cycles. . . ," thus tarnishing, for meteorologists, the image of Bacon.

6 Khrgian adds, "The hypothesis of solar effects on the weather played a considerable positive role in meteorology. It attracted the attention of many great scientists, and caused astronomers. . . . to become interested in meteorological problems and to gather a great deal of climatological material." The persistence of the effort to find a sunspot–weather correlation is remarkable. On 23 October 1987 *Science News* reported (p. 479), "After a centuries-long search . . . the first connection between variations of the sun and weather on Earth may have been found."

7 Economists have often postulated that economic behavior is cyclic. The economic cycles were sometimes believed to result from climatic cycles, as in Henry Ludwell Moore's *Economic Cycles: Their Law and Cause* (1914). William Stanley Jevons related financial crises to the sunspot cycle: "A mania is, in short, a kind of explosion of commercial folly, followed by the natural collapse. The difficulty is to explain why this collapse so often comes at interval of 10 or 11 years, and I feel sure the explanation will be found in the cessation of demand from India and China, occasioned by the failure of harvest there, ultimately due to changes in solar activity" (in Gregory, 1930, p. 114).

8 It was only in the late 19th century that reliable weather records for more than a decade or two were generally available. In 1873 R. H. Scott, Director of the Meteorological Office, complained (p. 380), "it is hardly possible to say what has been the approximate temperature of these islands for more than twenty years,—a period far too short for the definite recognition of a cycle."

9 The indirect testimony is given in Alter (1924, p. 483). Direct testimony to the tedium of computation is in Whipple (1924b, p. 239) and Shaw (1926, p. 277).

10 There were a number of highly complex machines that could have been immediately applied to the search for weather cycles: Kelvin's harmonic analyzer and Kelvin's tide predictor (1873), E. W. Blake's machine for drawing compound harmonic curves (1879), Henrici's harmonic analyzer (1894), Charles Chandler's "harmonograph" (1894), the Michelson-Stratton harmonic integrator (1898), Fischer and Harris's tide predictor (1911), the tide predictor of the Deutsches Hydrographisches Institut (1913), Hull's correlation-calculating machine (1925), and the continuous integraph of Bush, Gage, and Stewart (1927). These were built for other purposes, and in most cases only a single machine was built. In the 1890s Kelvin proposed that his harmonic analyzer be used to study the atmosphere, and this was apparently done (Shaw, 1934, p. 108). The others also may have been used in the search for weather cycles.

11 This information in this and the following two paragraphs comes from Burks and Burks (1988), Mauchly (1984), Stern (1980), and Tropp (1982).

12 The following are excerpts from Mauchly's testimony in the trial of Honeywell versus Sperry Rand: ". . . as soon as somebody told me that I had better quit there, stay out of this . . . that spurred me on"; "Now, here is something which the people at the Weather Bureau were sure did not occur. They were very much against the idea that the sun affected the surface weather, and it would take a preponderance of evidence to get them slightly interested in the idea that it might be true"; "And this was on a daily basis. So we could figure out—if you take 200 stations every day for 20 years, you've got a lot of data to add up even for just one weather variable" (in Burks and Burks, 1988, pp. 77, 79).

13 ENIAC was the first large-scale electronic digital computer; EDVAC was the first stored-program computer to be planned; BINAC was the first operational stored-program computer in the United States; and UNIVAC was the first commercially produced electronic digital computer (Stern, 1980).

14 Mauchly did present results orally at a meeting of the American Association for the Advancement of Science on 28 December 1940. (This talk, though of little significance for meteorology, had great significance for the development of computers, since it was at this session that Mauchly made the acquaintance of John V. Atanasoff.) The barrenness of his meteorological research and the fact that he was trained and initially employed as a physicist make it questionable to label Mauchly a meteorologist, as is sometimes done.

15 Gold adds, "The elimination of the consequent source of mistakes is even more important. . . ." Many meteorologists were attracted to machine computation more to avoid errors than to increase speed. (Similarly, many meteorologists were attracted to self-recording instruments more to avoid errors than to save labor.)

16 The other evidence cited by Miller was Richardson's book.

17 From Greek *νομος* (= law) and *γραφη* (= figure).

18 Both nomography and slide rules apparently met their quietus with the marketing of programmable pocket calculators in the 1970s. (The HP-35, introduced by Hewlett-Packard on 1 February 1972, was the first pocket calculator that computed trigonometric functions and logarithms.)

19 One might expect winds to blow from areas of high pressure to areas of low pressure. In fact, winds generally blow *along* isobars, their tendency to turn in the direction of lower pressure being restrained by the Coriolis and centrifugal forces. There is usually one speed (called the gradient

velocity) which results in an exact balance between the deflective tendencies. (The major movements of the solar system are analogous in that motions are generally perpendicular to the forces.)

In the equation given, ω is the angular speed of the earth, v is the wind speed, ϕ is the latitude, p is barometric pressure, n is horizontal distance perpendicular to the path in which the air is moving (so dp/dn is the pressure gradient), and r is the radius of curvature of the path. The equation may be read, "Coriolis force minus pressure-gradient force equals centrifugal force."

20 There is some argument in the meteorological literature about the adequacy of three-place accuracy; see Whipple (1924b). Further evidence that slide rules were routinely used in provided by the descriptions of the courses given in a 1-year training program for subprofessional positions in the Weather Bureau: the use of the slide rule was an important part of two courses [*BAMS* **32,** 109 (1951)].

21 In some countries daily forecasting involved more calculation. In the United States, as in most other countries, forecasts were almost entirely nonquantitative. But in France aviation forecasts included the expected heights of clouds, both upper and lower levels, and general forecasts, at least for Paris, included the expected high temperature and the expected wind speed. There must have been calculation to produce these numbers.

22 Shaw delighted in learned coinages ["I crave permission to use a new word "sistible" to describe a condition of possible statical equilibrium, derived from *sistere* just as stable is derived from *stare*. A stork standing on one leg is obviously *sistible* and by use of its muscles is also *stable*, a drunken man is really sistible but not stable . . ." (1926, p. 235)], and he was sensitive to a word's pedigree [calling "azimuth" "a convenient but rather uncouth Arabic word" (1928 p. xii)]. But it may be argued that he fell below his own standards in coining "tephigram," combining as it does Latin and Greek elements—"tee" and "phi" and "-gram." (The variables on the axes of a tephigram, often written θ and S, for temperature and entropy, were written by Shaw t and ϕ.)

23 Similar diagrams were in use in many countries. In Germany a diagram serving the same purpose was called the Neuhoff diagram.

24 In the 19th century, weather forecasts were sometimes called "probabilities," but there "probability" meant "probable weather." Probability theory played no role in the production of these forecasts, nor were numerical probabilities attached to them.

25 In the 1940s, and perhaps earlier, tabulating machines were used to produce evaluations of forecasts automatically (U.S. Department of the Air Force, Air Weather Service, 1948, p. 48).

26 Dobson defines it to be the difference between the error of the forecast and the actual change in the interval considered (Dobson, 1921, p. 266). Thus the most effective of the naive forecasting algorithms—predicting no change—gives a zero score when there is no change.

27 There are apparently only a few examples of atmospheric phenomena that were predicted by theorists before they were observed. The 1922 prediction of the thermal mesopeak by Lindemann and Dobson and the 1936 prediction of lee mountain waves by Queney are examples from the interwar period (Saltzman, 1967, p. 583).

28 The same year W. J. Humphreys in the United States gave a similar explanation of the temperature of the upper atmosphere.

29 In research practice there need not be a sharp distinction between theoretical investigations of the traditional sort and numerical experimentation. The availability of a computing staff or of an electronic computer sharpens the distinction, since one cannot turn work over to either without having specified an algorithm.

30 Ernest Gold said that "one of the most valuable features of the paper consisted in the tables" since such tables "simplified greatly the application of theoretical results to practical problems" (in Simpson, 1928b, p. 178).

31 Simpson's numerical experiments stimulated Elsasser also to develop some new graphical means of calculation (Rossby, 1957, p. 11).

32 The algorithm was presented in an appendix to the paper, just as, decades later, computer listings were often appended to papers.

33 The meteorologist of the 1920s who did most numerical experimentation, Richardson, was emphatic in his preference for numerical over graphical methods. In astronomy and physics, where some data were extremely precise, numerical methods were commonly used; and at the Edinburgh Mathematical Laboratory in the early 1920s, graphical methods had been all but abandoned (Richardson, 1924b, p. 163).

34 There is another way the facilitation of computation turns meteorologists away from meteorology: they become interested in the calculating devices themselves. This happened in this period with L. W. Pollak and John Mauchly; it happened in the 1940s with Maurice Wilkes.

Chapter 9

1 The main source of information for this account of the D-Day forecast is Stagg's book *Forecast for Overlord* (1971). A number of other meteorologists who took part in the forecasting have written about their experiences: Irving Krick in Krick and Fleming's *Sun, Sea and Sky* (1954); Werner Schwerdtfeger in "The last two years of Z-W-G [Zentral Wetterdienst Gruppe]" (1986); and a dozen or so others in *Some Meteorological Aspects of the D-Day Invasion of Europe 6 June 1944* (1986), edited by Shaw and Innes. There is additional information in Eisenhower's *Crusade in Europe* (1948) and other standard accounts. Weather charts for 3–6 June 1944 are retrospectively analyzed in Gold (1947).

2 In early March 1944 Overlord headquarters was moved from London, where Stagg had begun work the preceding November, to Teddington (a suburb southwest of the city), where it remained until the end of May. For the week before the invasion, headquarters was at Portsmouth (on the south coast). (Widewing is very close to Teddington, Dunstable is about 30 miles northwest of London, and Stanmore is a suburb northwest of London.)

3 This may have been a consequence of the fact that the German meteorologists had many fewer weather observations from the west than did the Allied meteorologists. Appendix A of Bates and Fuller (1986) is a reprinting of the weather report issued by the Saint-Germaine headquarters of Field Marshal von Rundstedt at 0500 hours on 5 June 1944.

4 It is indicative of the noticeability of meteorologists during World War II that also in the conquest of France 4 years earlier weather forecasting was widely believed to have contributed. It was thought that German meteorologists were especially adept at foreseeing "Hitler weather"—the dry ground and clear skies optimal for Blitzkrieg (assaults led by Panzer units supported by accurate artillery and dive-bombing Stukas) (Johnson, 1943, p. 201).

5 According to Krick and Fleming (1954, p. 174), these conditions were correctly forecast by the German meteorologists. A week after the beginning of this battle, General George Patton, commander of the U.S. Third Army, ordered his chaplain to distribute to the troops a prayer beginning "Almighty and most merciful Father, we humbly beseech Thee, of Thy great goodness, to restrain these immoderate rains with which we had to contend . . ." (in Bates and Fuller, 1986, p. 102).

6 "Weather stations on wheels" were first used in World War I (by the Germans); they were much used in World War II (Spilhaus, 1950, p. 362).

7 Each such weather station consisted of a ship, cruising in a circle of diameter about 10 miles, from which surface, pilot-balloon, and radiosonde observations were taken. [The Allied weathership

would leave its stations when, as often happened, a German submarine was reported in the area (H. L. Crutcher, personal communication, May 14, 1988)].

8 The destruction of that station was reported in the press as a significant victory (*New York Times*, Dec. 15, 1944, pp. 1, 3, 18). The U.S. Navy also developed automatic weather-observation buoys (Hughes, 1970, p. 169).

9 For example, in late 1943 weather task teams were parachuted into Yugoslavia in support of a planned bombing raid on the oil refineries of Ploesti, Romania (Hughes, 1970, p. 85).

10 *QJRMS*, **66,** 154 (1940).

11 Weather information being ubiquitous, censorship intruded in many activities. When a baseball game was halted because of rain, those listening to the broadcast were told that there would be a delay due to circumstances that the announcer could not name. When Eleanor Roosevelt mentioned in her newspaper column that rain had interfered with a lecture she had given, she was reprimanded by the Office of Censorship. However, in late 1943, as the threat of enemy air and submarine attack near the North American continent waned, many restrictions on weather information were relaxed (Whitnah, 1961, pp. 204, 205).

12 In World War I, the Russian meteorologist A. I. Voeikov argued for the utility of climatological information for the conduct of the war (Khrgian, 1959, p. 326).

13 The preface of a major American *Handbook of Meteorology,* published in 1945, contains the statement, "Much desirable material was omitted for security reasons at the time of writing" (Berry, 1945, p. vi).

14 The standard forms introduced were known as WBAN (Weather Bureau, Air Force, Navy) forms (Barger, 1960, p. 19).

15 In 1951 these groups moved to Asheville, North Carolina, where they have remained. (Now called the National Climatic Data Center, this center is the custodian of all U.S. weather records, receiving data from voluntary observers, the Coast Guard, and the Federal Aviation Administration as well as from the Weather Bureau, the Air Force, and the Navy.)

16 Sources of information on the use of punched-card equipment during World War II include *Machine Methods of Weather Statistics* (U.S. Department of the Air Force, Air Weather Service, 1948), prepared by the Air Force Data Control Unit of the Air Weather Service; *Climatology by Machine Methods for Naval Operations, Plans, and Research* (U.S. Department of the Navy, Aerology Branch, 1953), prepared for the Aerology Branch of the Office of Chief of Naval Operations; and Conrad and Pollak's *Methods in Climatology,* 2nd ed. (1950) (especially a letter from Reichelderfer reprinted on pages 351 to 353).

17 The deck was called the "Kopenhagener Schlüssel," perhaps from the place of capture.

18 This did not become standard practice in the U.S. Weather Bureau until November, 1975 (P. D. Thompson, personal communication, May 20, 1988).

19 An account of the analog method developed by Irving Krick, probably very similar to the one used at Widewing, is in Krick (1954, pp. 87–106).

20 The first large calculations on punched-card machines were not scientific calculations. They were done in the early 1920s by American railroad companies needing to compute car-mile and ton-mile statistics for the Interstate Commerce Commission (Eckert, 1940, p. ix).

21 Note April 14, 1945 by John Mauchly (Wexler Papers). Mauchly describes Hunt's computations as follows: "As part of a larger problem, it is necessary to know, on a number of days, the total amount of water vapor in the atmosphere above a given station. Pressure, temperature, and humidity readings have been made at various times at various altitudes (by captive balloon, I believe he said). Three or four thousand cards bearing these observations are prepared. But not all of these can be used, as on some dates the balloon did not gain sufficient altitude. Therefore, it is necessary

to sort out and reject many observations. About 400 cards remain to be worked with. Then the temperature readings must be converted to the absolute scale, and the reciprocal of this absolute temperature found. This must be multiplied by the pressure. Some correction factor for altitude must be applied, and a few further computations done, before the water vapor content of that particular stratum of the air is found. The cards must be then sorted once more, bringing all the cards for a given date together. Then the water vapor amounts for the various strata for a given time are added together, and the sum thus found is one of the desired results."

22 In the second half of the 19th century there was much interest in self-recording instruments. A particularly magnificent meteorograph was built in Rome in about 1860 by the famous astronomer Angelo Secchi. Secchi asserted, "All scholars are agreed on this point, that meteorology can only advance by having machines that record all the phenomena automatically" (in Middleton, 1969, p. 260, see also Scott, 1869). The development of electronic computers in the 1940s prompted the first proposal for an automatic device to take observations *and* make forecasts. In 1950 Athelstan Spilhaus wrote (p. 363), "We are beginning to have the complete set of components with which to build the ultimate weather instrument. . . . [it] will observe at all suitable points in three dimensions and transmit to a center. There it will not only store data and compile it for study purposes, but will pick off and pre-compute the information needed to feed the electronic computing devices which will prepare predictions and automatically disseminate them to the distribution networks."

23 It became a popular science too. Quite a few popular books on meteorology were published during the war, and the widespread interest in meteorology among British servicemen was shown by the great demand for lectures by members of the Royal Meteorological Society [see *QJRMS,* **66,** 363–364 (1940)] and by the meteorological activities of British prisoners-of-war in Germany [see *QJRMS,* **69,** 182 (1943) and J. Weston (1945), which tells something also of the suspicion attached to such activities in wartime.]

24 In an article on the history of computing Garrett Birkhoff (1980, p. 23) says: "I think that in historical honesty we have to realize that it was dedication to the struggle against Hitlerism, and later to other problems of national defense, that provided the main driving force behind the development of the computer in the 1940s. It's absolutely impossible to understand it except in that context."

25 Byers (1970, p. 215) wrote that all together 7000 meteorological officers were trained in the United States during the war.

26 The training manual, issued by the Navy's Bureau of Aeronautics, was "intended as a text and *workbook*" (p. iii, emphasis added). Other examples are Bernard Haurwitz's *Dynamic Meteorology* (1941) and Charles Barber's *Weather Science* (1943).

27 Almost all algorithms in meteorology were mathematical, and almost all the mathematical parts of meteorology could be used to generate exercises (pedagogical algorithms) even if they were not the basis of algorithms used in research or forecasting (practical algorithms). (The assignment of a number to wind force according to the Beaufort scale is an example of a nonmathematical algorithm, the carrying out of which requires observational rather than mathematical skill.)

28 *BAMS* **32,** 107 (1951).

29 Quite a few other colleges and universities offered some training in meteorology, but did not have autonomous departments.(Thompson, 1987, p. 635). Although MIT established a professional program in meteorology in 1928, meteorology was not a separate department there until 1940. Beno Gutenberg started a department of meteorology at Cal Tech in 1934. NYU's Department of Meteorology was started by Athelstan Spilhaus in 1938. The Institute of Meteorology at the University of Chicago was established in 1940 by Rossby and Byers. The German takeover of Norway left Jacob Bjerknes stranded in the United States; at Rossby's urging, UCLA in 1940 set up a department of meteorology around Bjerknes. [Two other Norwegian meteorologists stranded in the United States were H. U. Sverdrup and Sverre Petterssen (Byers, 1970, p. 214).]

30 Cal Tech discontinued its meteorology department; Cornell, Florida State, Oklahoma A&M, Penn State, University of Utah, University of Washington, and University of Wisconsin established meteorology departments.

31 The *Journal of Meteorology* is the current *Journal of the Atmospheric Sciences*, the name having been changed in 1962.

32 The main purpose of the Field Information Agency, Technical (FIAT) review of German meteorology from 1939 to 1946 (Mügge, 1948) was to make wartime research known (within Germany as well as internationally). Although the *Quarterly Journal of the Royal Meteorological Society* continued to publish research results during the war, the referees of each of the submitted articles were asked whether the article contained information that might be of value to the enemy, in which case publication could not occur until the war was over [*QJRMS* **66,** 2 (1940)].

33 Radiosonde observations were called RAOBs, airplane observations, APOBs.

34 Recall (from Chapter 8) that on a tephigram (or the German near-equivalent, a Neuhoff diagram) one plots observed values of entropy and temperature at various heights. The plot can then be used to answer a variety of questions, such as whether an air column is stable or whether clouds will form at a particular height. In wartime it was often an important question whether condensation trails would form behind an aircraft, and this question, too, was answered by using the tephigram.

35 Wobus's device was known as the "wiggle wagon."

36 Military-surplus radar was put to meteorological use also in England (Hitschfeld, 1986, p. 33).

37 The use of the captured V-2s is described in DeVorkin (1987).

38 But not because of any use of atomic explosions to control the weather, something several scientists thought might be possible. The fact that von Neumann did work in computational fluid dynamics for the Manhattan Project may have influenced his choice of weather forecasting as the major problem to be studied as part of his postwar computer project at the Institute for Advanced Study (see Chapter 10). From the beginning, meteorologists were involved in the testing of the atomic bomb, being asked to predict atmospheric transport of radioactive particles. And in the past decade much activity in atmospheric modeling has been connected with the "nuclear winter" debate.

39 In 1970 Edward Lorenz wrote (1970a, p. 24), "To the best of my knowledge every forecast of the state of meteorology today, made prior to World War II, has been more or less a failure, and the one dominating reason for this failure is failure to forecast the invention and development of the computer."

40 This was clearly the case with John von Neumann's computer project at the Institute for Advanced Study. Meteorology came to be mentioned often as a field in which a new computer might be used to advantage. For example, in a lecture entitled "A review of government requirements and activities in the field of automatic computing machinery," which was one of the famous Moore School Lectures given in August 1946, J. H. Curtiss talked quite a bit about the possible utility of computers in meteorology (Campbell-Kelly and Williams, 1985, pp. 349–350).

41 Although Simpson started the modern study of the radiation balance, the idea was not new. In about 1880 the Russian meteorologist A. I. Voeykov wrote that one of the most important investigations was the "introduction of an input–output table of solar heat received by the earth, with its spheres of air and water" (in Fedoseev, 1976, p. 53).

42 Starr later wrote that he was "Encouraged by the later work of Rossby and the suggestions of the geophysicist Harold Jeffreys" to undertake "an extensive and protracted program of deriving from massive amounts of data the actual dynamics of the large-scale processes of the atmosphere" (Starr, 1980, p. 148). Lorenz has written (1983, p. 733), "Following World War II, Bjerknes (1948), Priestly (1949), and Starr (1948) independently proposed that upper-level observations had

now become plentiful enough for the direct evaluation of transports of angular momentum on a day-by-day basis."

43 There was even some success at weather modification with the program known as FIDO (Fog, Investigations Dispersal Of) (Hughes, 1970, p. 183).

44 In 1941 Reichelderfer wrote (p. 150), "In the present state of meteorology an explicit forecast in the accepted sense of the term is not usually justifiable from the scientific viewpoint for longer than 48 hours beyond the time of issue." In 1944 Douglas said that "the production of regular forecasts which have any true scientific validity or worthwhile dependability is likely to be out of the question for more than two days ahead, or even, at times, more than 24 hours" (in Stagg, 1971, p. 38).

45 This project was sponsored by the Bureau of Agricultural Economics and the Weather Bureau (at that time part of the Department of Agriculture) in the hope that long-range forecasting could prevent much suffering of the sort being felt in the mid 1930s by Midwestern farmers struck by drought.

46 It is interesting that apparently the only forecaster in the group at Dunstable who placed much faith in longer-range forecasting was Petterssen, who, although he had practiced as a forecaster in Norway and the United States, was "Untrammelled with the humbling experience of practising in England" (Stagg, 1971, p. 54). (It may be argued that the complexity of the stratigraphy in the British Isles was an inducement to the development of English geology. Similarly, the difficulty of predicting the weather in England may have stimulated English meteorology; it did, certainly, make English meteorologists skeptical.)

47 The war stimulated interest in long-range forecasting also in Germany. Before the war Fritz Baur, who in 1930 had established an institute near Frankfurt that regularly issued 5- and 10-day forecasts, was shunned by most of the meteorological community because long-range forecasting was considered unscientific. During the war his institute was expanded and taken over by the German air ministry, and Baur was ordered to distribute his forecasts to about 25 military officers (see Landsberg, 1978; and Neumann and Flohn, 1987).

48 In 1977 Richard Reed wrote, "From such records as do exist it appears that only a slight increase in forecast accuracy was achieved during the 1860–1920 era" and "In principle the introduction of frontal analysis and later of upper air analysis during the 1920–50 era should have resulted in some increase in predictive ability, and likely it did. However, the increase must have been small, since it is hard to find factual support for this belief" (p. 395). In 1946 H. G. Houghton, President of the American Meteorological Society, said, "Such comparisons as have been made indicate that there has been no significant increase in the accuracy of short-range forecasts in the past 30 or 40 years" (in Douglas, 1952, p. 16).

49 In about 1880 a German meteorologist reported that those who were least content with the official weather predictions were the forecasters themselves (Scott, 1883, p. 73).

50 The Office of Naval Research, established 1 August 1946, administered much of the research funding for meteorology [*BAMS* **67,** 1414—1415 (1986)].

51 *BAMS* **32,** 107 (1951).

52 *BAMS* **32,** 104 (1951).

53 Throughout the war and afterward, cooperation among the civilian and military weather services was promoted by the Joint Meteorological Committee of the Joint Chiefs of Staff.

54 Each major military unit had its own meteorological staff because of the felt need for "direct personal contact between the meteorological expert and those responsible for planning, directing and executing operations" (Johnson, 1943, p. 203).

55 Speaking in 1952 of forecasting in general, C. K. M. Douglas said that the amount of subjectivity "is generally recognized to be appreciable, though on the average a good deal less than the difference between forecast and actual events" (p. 17).

56 The military meteorologists who did have extensive experience were in many cases assigned to a place having weather quite unlike what they were familiar with. The synoptic methods most useful in one region could be quite different from those useful in another (Shaw, 1932, p. 396).

57 Chief of Naval Operations 1967 p. 2–1. Navy meteorologists were, for example, unable to agree about the path of the December 1944 typhoon, which caused the loss of 790 men and several ships (Krick and Fleming, 1954, p. 169).

58 Note April 14, 1945 by John Mauchly (Wexler Papers).

59 The method was considered unsafe by M. G. Kendall (Conrad and Pollak, 1950, p. 425).

60 The decline after 1950 in the number of articles on objective forecasting is seen in the graph included in Gringorten's 1953 study referred to earlier.

Chapter 10

1 Memorandum Wexler to Reichelderfer 20 Dec 46 (Wexler Papers). The full passage is as follows: "II. Numerical Forecasting. This type of forecasting, based on a "brutal assault" of the equations of motion, has been attempted by Richardson, Elliott, and Thompson. Professor Panofsky is also giving this matter some study. Both Thompson and Panofsky agreed, and the conferees concurred, that the time is not ripe for such a numerical forecast system, principally because the geostrophic terms in the equations of motion are so very large compared to the remaining terms whose effect, however, is still very significant in changing the weather patterns."

2 Letter Rossby to Reichelderfer 16 Apr 46 (von Neumann Papers). On several occasions von Neumann contrasted the success of mathematics in handling the linear differential equations of electrodynamics and quantum mechanics with the failure of mathematics in handling the nonlinear differential equations of hydrodynamics and expressed his belief that in the latter case computer-aided numerical experimentation would lead to mathematical, not merely practical, advance. [See *Collected Works of John von Neumann,* Vol. 5, pp. 2–3 (1963a) and Vol. 6, pp. 357–359 (1963b); and Charney, 1972, p. 117.]

3 The acronyms come from "Electronic Numerical Integrator and Computer" and "Electronic Discrete Variable Arithmetic Computer."

4 As it turned out, most of the funding for the computer project at the Institute came from various agencies of the U.S. government: the Office of Naval Research, Army Ordnance, a tri-service contract, and the Atomic Energy Commission (Goldstine, 1972, p. 243).

5 Letter von Neumann to Brown 28 Nov 45 (von Neumann Papers). There is significant overlap of fluid dynamics and meteorology, but the former is largely a laboratory science (as are the other sciences here listed by von Neumann), unlike meteorology.

6 Aspray (1990, pp. 130–132) considers the possibilities at greater length.

7 In early 1945 John Mauchly had discussed with people at the Weather Bureau the possibility of using the EDVAC in meteorological research and weather prediction.

8 Zworykin, who was born in Russia in 1889 and emigrated to the United States in 1919, is famous as the inventor of the television camera and the television receiver. He also invented the scintillation counter and played an important part in the invention of the electron microscope. He may have become interested in weather prediction as a result of the development of meteorological instruments at RCA.

9 Supporting paper "Weather proposal by V. K. Zworykin" 30 Jan 46 (Wexler Papers). Zworykin was interested also in forecasting by means of electronic analog-simulation of the weather (Charney, 1987, p. 54).

10 See Letter Reichelderfer to Zworykin 29 Dec 45 (Plaintiff's Trial Exhibit No. 3867.2, Honeywell vs Sperry–Rand, Babbage Institute Archives).

11 Five days later another meeting was held, presumably also in Washington, this one attended by Mauchly, Eckert, and representatives of the Army, Navy, and Weather Bureau (Supporting Paper cited above). Mauchly and Eckert were primarily interested in using computers for the analysis of climatological data. [See also Letter Reichelderfer to Elliott 14 May 46 (Wexler Papers).]

12 The article implied that von Neumann and Zworykin alone had conceived of the "super calculator": "Dr. John von Neumann . . . and Dr. Vladimir Kosma Zworykin . . . developed the plans for the new machine."

13 The War Department released information about the ENIAC and the EDVAC about a month later; this was reported on the front page of the *New York Times* on 15 February.

14 It was probably this article that figures in the story of how Philip Thompson, a young Air Force meteorologist, joined von Neumann's project late in 1946 (see Thompson, 1983, p. 758).

The meteorological community seems to have ignored the claims that the computer might lead to control of the weather. In von Neumann's proposal to establish the Meteorology Project [Letter Aydelotte to Rex 8 May 46 (Charney Papers)] is the claim that, as a result of the Project, "the first step towards influencing the weather by rational, human intervention will have been made—since the effects of any hypothetical intervention will have become calculable." Yet weather control does not seem to be mentioned in any of the subsequent Project letters and reports, with the exception of a trip report from Wexler to Reichelderfer on 18 October 1946 (Wexler Papers). Wexler had had dinner with Zworykin and reported, "Dr. Zworykin is still interested in weather control and, as a result of some recent visual observations he made off Miami Beach of the build-up of cumulonimbus . . . he thinks that hurricanes might be prevented from forming by igniting oil on the sea surface in critical areas, thus "bleeding off" energy by thunderstorms which might otherwise go into hurricane formation."

15 "It was largely the great regard in which von Neumann was held which allowed him to obtain the funding for his [computer] project" (Williams, 1985, p. 353).

16 The most dramatic example is his arranging, by a single phone call to a superior officer, the transfer of Thompson to Princeton (Thompson, 1983, p. 758).

17 Letter Rossby to Reichelderfer 16 Apr 46 (Wexler Papers), and Thompson (1986, p. 6).

18 Memorandum Wexler to Reichelderfer 12 Sep 46 (Wexler Papers).

19 For an example of a task von Neumann set for the meteorologists, see Memorandum Wexler to Reichelderfer 20 Dec 46 (Wexler Papers).

20 See Letter Elliott to Reichelderfer 4 June 1946 (Wexler Papers).

21 The quotations are from Elliott's typewritten manuscript "Calculated weather predictions," which was written August 1943 and condensed September 1946. A copy of this manuscript is in the Wexler Papers.

22 Letter Rossby to Reichelderfer 16 Apr 46 (von Neumann Papers).

23 See Letter von Neumann to Queney 18 May 46 (Wexler Papers), and Letter Aydelotte to Rex 8 May 46 (Charney Papers). The proposal is signed by Frank Aydelotte, Director of the Institute for Advanced Study, but it was probably written almost entirely by von Neumann.

24 In fact it took almost 5 years to complete the computer, difficulties in building the primary memory device accounting for most of the delay.

25 Letter Wexler to Forsythe 22 Jul 46 (Wexler Papers).

26 Memorandum Wexler to Reichelderfer 12 Sep 46 (Wexler Papers). See also Progress Report 1 Jul 46 to 15 Nov 46 on Contract N6ori-139 (Charney Papers); this is the first of a full set of progress reports for the Meteorology Project included in the Charney Papers.

27 Especially important, as things turned out, was that this meeting aroused Charney's interest in numerical prediction (Charney, 1987, p. 85).

28 Memorandum Wexler to Reichelderfer 12 Sep 46 (Wexler Papers).

29 The office space and housing shortages were soon overcome. The Meteorology Project got office space in the Institute's new computer building, which was completed in January 1947. And according to the Progress Report 1 Jul 46 to 15 Nov 46, "With regard to the housing situation, the Institute has made an exceptional effort, by acquiring 13 houses equalling 32 family units, from a Government settlement at Mineville, New York, transporting them by rail to Princeton, and re-erecting them here, all at its own expense." Thompson, Pekeris, and Goldstine were, by April, living in such units (Progress Report 15 Nov 46 to 1 Apr 47). (Herman Goldstine, although not a meteorologist, was a participant in the Project in the capacity of numerical analyst.)

30 Memorandum Wexler to Reichelderfer 4 Nov 46 (Wexler Papers).

31 Progress Report 15 Nov 46 to 1 Apr 47. In connection with this study von Neumann arranged to use the SSEC (Selective Sequence Electronic Calculator) at IBM headquarters in New York.

32 Progress Report 15 Nov 46 to 1 Apr 47.

33 Memorandum Wexler to Reichelderfer 20 Dec 46 (Wexler Papers).

34 According to Progress Report 15 Nov 46 to 1 Apr 47, "Resulting representations are found to agree better with subjective synoptic analyses than do the individual subjective analyses agree among themselves." In the same document, von Neumann reported that the Mathematical Tables Project "has been very helpful in various other numerical tasks as well."

35 Progress Report 15 Dec 47 to 15 May 48.

36 See Charney (1987, pp. 21–24) for Charney's account of his change of fields. Charney's happening to hear a lecture by the meteorologist Jörgen Holmboe and a conversation with the fluid dynamicist Theodore von Karman were important in the decision to change fields. (Holmboe later served as Charney's thesis advisor, although Charney reported that he received little guidance from Holmboe, partly because Holmboe was entirely geometric and visual in his thinking while Charney was quite algebraic and analytic.)

37 In a 1947 letter to Thompson, Charney vividly describes the variety of motions described by the general equations: "In the terminology which you graciously ascribe to me we might say that the atmosphere is a musical instrument on which one can play many tunes. High notes are sound waves, low notes are long inertial waves, and nature is a musician more of the Beethoven than of the Chopin type. He much prefers the low notes and only occasionally plays arpeggios in the treble and then only with a light hand. The oceans and the continents are the elephants in Saint-Saens' animal suite, marching in a slow cumbrous rhythm, one step every day or so. Of course, there are overtones: sound waves, billow clouds (gravity waves), inertial oscillations, etc., but these are unimportant and are heard only at N.Y.U. and M.I.T." (in Thompson, 1983, p. 767).

38 It is not difficult to eliminate insignificant solutions. For example, one can treat the atmosphere as two-dimensional by ignoring vertical motions and thus ban gravity waves from the model. But by so doing one would ban as well types of atmospheric behavior, such as the generation of storms, considered meteorologically significant. The meaning of "significant" is, of course, variable. If Rossby waves are the only type of large-scale motion considered significant, then Rossby in 1939 had an appropriate filter (Thompson, 1952, p. 33).

39 Arnt Eliassen, with whom Charney worked at Oslo, arrived at the same equation from a quite different starting point (see Platzman, 1979, p. 307).

40 The time increment must be less than the time required for a wave impulse to be transmitted from one grid point to another (Thompson, 1983, p. 143).

41 Since Richardson did not go beyond one time step, the failure of his predictions was not the result of computational instability.

42 The work was published in 1948 under the title "On the scale of atmospheric motions." Thompson called this article "the most significant contribution to numerical weather prediction since Richardson's magnum opus" (Thompson, 1983, p. 759).

43 Letter Charney to von Neumann 2 Jan 48 (von Neumann Papers).

44 Namias was one of those invited to make regular visits to Princeton in order to increase the input to the Project from empirical meteorology.

45 Charney explained, "By introducing new physical factors one at a time into these numerical experiments one has a control over their effect; if they were introduced all together it would be impossible to ascertain the factor or factors responsible for the discrepancies. . . . The primary reason for Richardson's failure may be attributed to his attempt to do too much too soon" (Progress Report 1 Jul 48 to 30 Jun 49).

46 Phillips gives on p. 495 an illuminating example from Charney's later work of Charney's insistence on comparison with the real world.

47 Progress Report 1 Jul 48 to 30 Jun 49.

48 This height was taken to be that at which the pressure is 500 millibars. (Sea-level pressure is about 1000 mb.)

49 Even here some skill at numerical analysis was necessary. On p. 41 of the published report the authors say, "The series (13) for the influence function converges rather slowly, so that the straightforward computation by evaluating a sufficient number of terms in the series would be a very laborious task. However, the authors' associate, **G. Hunt,** has given a method by which the computation can be done without too much work."

50 Progress Report 1 Jul 48 to 30 Jun 49.

51 Progress Report 1 Jul 48 to 30 Jun 49.

52 This view is expressed in Progress Report 1 Jul 48 to 30 Jun 49, in Charney and Eliassen (1949) and in Sawyer and Bushby (1951).

53 See Progress Report 1 Jul 49 to 30 Jun 50 (Smagorinsky, 1983, p. 9), and Letter Reichelderfer to Hughes 22 Sep 49 (von Neumann Papers).

54 ENIAC was built as a hard-wired machine, which is to say that to change the algorithm carried out by the machine, one had to change the wiring. Von Neumann realized that ENIAC could, without a great many changes, be converted to a stored-program machine. This conversion was made in 1948, more than a year before the meteorology group made use of the machine (Platzman, 1979, p. 301).

55 Progress Report 1 Jul 49 to 30 Jun 50.

56 Minutes of the meeting held 5 August 1952 at the Institute for Advanced Study (Wexler Papers).

57 Platzman (1979, p. 311) wrote: "Of all the difficulties that plagued us, however, by far the most baffling and certainly the most disconcerting was the assignment of scale factors for the individual ENIAC operations. These factors intruded because ENIAC was strictly a fixed-point machine, each register holding a fixed-point decimal number with 10 digits and sign. The purpose of the scale factors was to prevent overflow or underflow or simply excessive loss of significance in each of the various ENIAC operations."

58 More precisely, to make "forecasts from a theoretical model of stable and unstable waves on zonal jet-like flows in a channel bounded by streamlines on two latitude circles" (Platzman, 1979, p. 303).

59 This type of instability manifests itself in the change of a smooth flow into a large number of jets—Charney called the process "noodling." (My information on the second ENIAC expedition comes mainly from a conversation with George Platzman on 20 December 1987.)

60 The fractional dimension is explained by the following convention: if the vertical dimension is represented by n layers, then the model is said to have dimension $2 + (n - 1)/n$.

61 Limited operation began at the end of 1951; public announcement of its operation was made in June 1952 (Pugh, 1984, p. 17).

62 The *New York Times* reported on Monday 27 November 1950, "New York and most of the northeastern states emerged yesterday from the great storm that had battered them on Saturday to count 205 dead and property damage estimated at more than $100,000,000. . . . The great and unexpected storm [was] described by the Weather Bureau here as "the most violent of its kind ever recorded in the northeastern quarter of the United States". . . ."

63 Progress Report 1 Jul 52 to 30 Sep 52.

Chapter 11

1 The following appeared in an article written in 1952 by Staff Members of the Institute of Meteorology of the University of Stockholm (p. 21): "During the last few years several groups of meteorological research workers have devoted an increasing amount of effort to the problem of translating basic dynamic principles into practically usable methods for quantitative objective forecasting . . . . The advent of high-speed electronic computing machines has been a factor of decisive importance in this development . . . ." And George Platzman, in describing the first ENIAC expedition, wrote (1979 p. 307), "On a contracted time scale the groundwork for this event had been laid in Princeton in a mere two to three years, but in another sense what took place was the enactment of a vision foretold by L. F. Richardson 50 years before."

2 Charney attributed this change to the development of electronic computers. It is also significant that in 1965 Richardson's *Weather Prediction by Numerical Process* was republished and reviewed in major meteorological journals.

3 Particularities of a computer could be constraining in other ways; for example, there were the difficulties, mentioned earlier, in setting the scale factors for the ENIAC computations.

4 An example may be taken from the Progress Report for 1 July 1952 to 30 September 1952 written by Norman Phillips: "To aid in diagnosing malfunctions of the machine, a form of the barotropic code has been rewritten by Glen Lewis to check all of the major arithmetical operations which occur in the normal form of the barotropic code, and has proved quite useful."

5 In the same report (p. 238) the slowness of the ENIAC was one reason given for not attempting the integration of the Eulerian equations.

6 Letter Starr to von Neumann 29 Oct 47 and Letter von Neumann to Starr 4 Nov 47 (von Neumann Papers).

7 Progress Report 1 Oct 51 to 21 Dec 51.

8 Progress Report 1 Jul 51 to 30 Sep 51.

9 Minutes of the meeting held 5 August 1952 at the Institute for Advanced Study (Wexler Papers), p. 3. At this meeting von Neumann said, "I believe best results here would be obtained by attempting to divert prospective physicists and applied mathematicians into meteorology" (p. 7).

10 Letter Charney to von Neumann 2 Jan 48 (von Neumann Papers).

11 Current terminology is somewhat different: "short-term forecasting" for 1 to 48 hours, "medium-term" for 2 to 14 days, and "long-term" for more than 14 days.

12 Letter Reichelderfer to Taber 10 Apr 50 (von Neumann Papers).

13 Minutes of a meeting held 5 Aug 52 at the Institute for Advanced Study (Wexler Papers). All of the quotations in the next two paragraphs are from these minutes.

14 Many meteorologists were, of course, trained in mathematical physics, but essential to devising computer algorithms was skill in numerical analysis, and it was this sort of mathematics von Neumann had in mind when he said that "the meteorological and mathematical aspects probably [could not] be separated."

15 It appears that they arrived on about 1 January 1953 (Summary of Work, calendar year 1952, p. 6, and Progress Report 1 Jan 53 to 31 Mar 53). The Air Force and Navy representatives remained for one year, the Weather Bureau representative longer.

16 The Weather Bureau sent five meteorologists to an intensive (and probably intense) 10-week course that Platzman offered in the summer of 1953 (Smagorinsky, 1983, p. 22).

17 Memorandum Wexler to Reichelderfer 31 Jul 53 (Wexler Papers).

18 The Ad Hoc Committee had recommended the rental of an IBM 701. The Joint Meteorological Committee asked for a consideration of the Remington Rand 1103 [also known as the ERA (= Engineering Research Associates) 1103]. In January 1954 Herman Goldstine and Joseph Smagorinsky reported on the competitive tests they supervised. They found that in speed the machines were comparable, but that the 701 was significantly faster in input and output. (See Goldstine, 1972, p. 329. A copy of the original report is in the Wexler Papers.)

19 Letter Rex to Alt 15 Oct 53 (Wexler Papers).

20 Wexler, "Proposal for a project on the dynamics of the general circulation" 1 Aug 55 (von Neumann Papers).

21 The main participants were three Swedes—B. Bolin, G. Dahlquist, and B. Döös —two Icelanders—G. Arnason and P. Bergthorson—and two Americans—H. Bedient and N. Phillips. The group received financial support from the Wallenberg Foundation, the U.S. Office of Naval Research, the Swedish National Science Research Council, and the U.S. Weather Bureau (Staff Members, Institute of Meteorology, University of Stockholm, 1954, p. 140).

22 This average was only slightly higher than that reported by Bolin and Charney 3 years earlier (1951) and that reported by the Stockholm group 2 years earlier (Staff Members, 1952), but it was considerably higher than that reported by Sawyer and Bushby (1951). The earlier trials were, of course, not done on a real-time basis.

23 Proposal for a Project on the Dynamics of the General Circulation 1 Aug 55 (von Neumann Papers).

24 In October 1955 he sent a questionnaire to meteorologists at almost all national weather services and to meteorologists at the major research and academic institutions worldwide. With the exception of the meteorologists in the Soviet Union, response to the questionnaire was fairly complete. The results of the survey were summarized in a mimeographed report prepared by Gene Birchfield, which was sent in June 1956 to all those who had returned the questionnaire. A copy of the questionnaire, all the replies, and the summary have been placed in the archives of the Charles Babbage Institute. Most of the information in this and the following paragraph comes from the Platzman survey.

25 The General Circulation Research Section was headed by Joseph Smagorinsky. Jule Charney directed the work at Princeton, and W. Lawrence Gates the work of the Atmospheric Analysis Laboratory. At Chicago H. Riehl and C. Jordan were in charge of the work on numerical forecasting; at Los Angeles, J. Holmboe and M. Holl. The work at Uccle, Belgium, was directed by J. Van Mieghem; the work at Frankfurt am Main, by K. H. Hinkelmann; and the work at Dunstable, by J. S. Sawyer.

26 W. L. Godson was directing the Canadian effort, and C. L. Godske the work at Bergen. The Meteorological and Hydrographical Institute was headed by A. Nyberg, and the Norwegian Meteoro-

logical Institute, by R. Fjörtoft. E. Chararsch, of the Meteorological Service of Israel, was to be in charge of research in numerical forecasting and was at the time of the survey studying at the Institute of Meteorology in Stockholm. The work in London was directed by E. T. Eady, at Woods Hole, by J. S. Malkus, and at MIT, by V. P. Starr.

27 Fjörtoft's group was waiting for the completion of a Norwegian computer. M. Kwizak (1972, p. 1155) explained the Canadian slowness in beginning work with numerical weather prediction as follows: "First it appeared that only countries with large resources could afford the luxury of expensive computing machines, especially for real-time operational programs, and second, there was no computer available to the group in Montreal for research and experimentation in NWP techniques."

28 In the late 1950s Philip Thompson wrote, "The limiting factor [in the development of dynamical methods in the Soviet Union] is the availability of a high-speed automatic computer at least comparable with the IBM 704. Evidently no such machine is currently available in the USSR for the exclusive use of weather forecasting" (in Baum and Thompson, 1959, p. 408).

29 This information comes mainly from Cressman (1965) and from the national reports on numerical weather prediction that were distributed at the Symposium on Numerical Weather Prediction held in Tokyo in November 1968 (copies of which are in the library of the National Climatic Data Center in Asheville, NC).

30 In 1978 Fleet Numerical Weather Central became Fleet Numerical Oceanography Center.

31 Human forecasters did succeed, in the trials made in 1961, in improving on the the computer-generated 500-mb forecasts, but the "improvements were not sufficient to make up for the decay of forecast during the concomitant delay."

32 Computer graphics have been important in alleviating the problem identified by John Bellamy in 1952 in a review article entitled "Automatic processing of geophysical data": "The greatest shortcoming is the inability of the automatic equipment to produce output records which are capable of convenient and efficient mental assimilation and interpretation" (p. 42).

33 It has been customary to distinguish several "generations" of computers: computers of the first generation used a great many vacuum tubes, whereas those of the second generation used mainly transistors; computers of the third generation incorporated integrated circuits, and those of the fourth generation large-scale integrated circuits. For a discussion of the terminology of computer generations see Aspray (1985).

34 Much of the information in this paragraph comes from Williams (1985).

35 These improvements would not have come so rapidly without the "commercialization" of the computer. In 1950 computers were seen mainly as research instruments, to be designed and built at universities and other research institutions; few people thought of them as a commercial product. By the late 1960s hundreds of different computers were being marketed by some 80 major manufacturers (Eames and Eames, 1973, p. 163).

36 The new models produced 500-mb forecasts that, for the first time, were clearly superior to subjective forecasts, and, as a result, "the computer-produced 500-mb chart was substituted for the subjectively prepared 500-mb chart on all domestic facsimile weather circuits" (Bates and Fuller, 1986, p. 151).

37 The cost of a large system in this period was about $1.5 million (Wolff and Hubert, 1964, p. 643). The estimates of computer speed given above, from Bates and Fuller (1986, p. 188) and Wolff and Hubert (1964, p. 642), were intended to apply to meteorological computations. Wolff took as typical 10 additions per multiplication.

38 The earlier memory of Whirlwind was not in fact based on the Williams tubes, but on a similar cathode ray tube device (Williams, 1985, p. 322).

39 *BAMS* **44,** 98 (1963).

40 Other national meteorological services that soon acquired an IBM 360 were the U.S., the Belgian, and the Italian.

41 The importance of computer technology to the sister discipline of oceanography is also clear: the great interest in the 1980s in numerical modeling of the ocean has been explained as a result of the availability for the first time of computers powerful enough to implement realistic models of the ocean (Tribbia and Anthes, 1987, p. 498).

42 Of course there continued to be problems: in 1972 George Cressman wrote (p. 199) that malfunction of hardware was a major problem at the National Weather Service.

43 An important issue, which is for the most part outside the scope of this book, is the precise relation between computer capabilities and meteorological models. In 1978 Ronald Hughes, commanding officer of the Navy's Fleet Numerical Weather Central, testified before a committee of the House of Representatives that "the two most important investments this nation can make to accelerate progress in weather forecasting are in satellite technology development and increased computer power for the three numerical weather prediction centers" (U.S. Congress, 1978, p. 7). Hughes added that "modern weather forecasting is a partnership between man and computer."

44 In the mid 1980s a computer, the Navier-Stokes Computer, was designed specifically for solving problems in fluid mechanics, and modeling of the atmosphere was one of its intended applications.

45 Von Neumann's principal meteorological interest was in numerical forecasting. Both Mauchly and Wilkes were interested in the effect of the sun on the atmosphere; Mauchly, as we saw in Chapter 8, hoped to find evidence of an effect on the weather, and Wilkes made numerical investigations of atmospheric tides. Wilkes wrote (1985, p. 115): "Atmospheric oscillations became a major interest which I pursued for a good many years in parallel with the work on digital computers. . . . Since computation played a part in the study of atmospheric oscillations both for the evaluation of theoretical models and for the reduction of barometric and geomagnetic observations, I was able, to a certain extent, to combine both interests."

46 John Backus, one of the designers of Fortran, wrote the following about programming in the 1950s: "The programmer had to be a resourceful inventor to adapt his problem to the idiosyncrasies of the computer: He had to fit his program and data into a tiny store, and overcome bizarre difficulties in getting information in and out of it, all while using a limited and often peculiar set of instructions. He had to employ every trick he could think of to make a program run at a speed that would justify the large cost of running it. And he had to do all of this by his own ingenuity, for the only information he had was a problem and a machine manual" (Backus, 1980, p. 125).

47 Of course tables are still a standard way of presenting observational data. Tables have served also to present "calculated data," as with Edmond Halley's 1687 calculations of barometric pressure as a function of height, or with K. L. Coulson, J. Dave, and Z. Sekera's 1960 *Tables Related to Radiation Emerging from a Planetary Atmosphere with Rayleigh Scattering* (prepared with the use of an electronic computer). Such tables may be regarded as calculating aids, since they serve to obviate a lengthy calculation rather than to present empirical findings.

48 In 1981 the Board of Certified Consulting Meteorologists of the American Meteorological Society endorsed a list, prepared by Chester Newton, of 50 books useful to a consulting meteorologist; 2 of the 50 books are compilations of tables (Newton, 1981a, pp. 71–77).

49 The examples named are described in, respectively, Riehl (1952), Weiss (1955), Peterson (1961), and *BAMS,* **45,** 542 (1964).

50 Particularly clear evidence that nomograms were giving way to computers can been seen in a 1960 article in the *Bulletin of the American Meteorological Society* by Earl E. Lackey entitled "A method for assessing hourly temperature probabilities from limited weather records." Lackey presents a nomogram, then shows how the same calculation could instead be carried out by computer.

51 These were the Air Weather Service of the Air Force, the Extended Forecasting Section of the Weather Bureau, the Travelers Weather Research Center in Hartford, Connecticut, and by depart-

ments of meteorology at MIT, the University of Chicago, Pennsylvania State University, and the University of Washington. At the same time or only slightly later, Fjörtoft's method was being used by the Canadian Meteorological Centre (Kwizak, 1972, p. 1155).

52 Letter Godske to Platzman 24 Oct 55 (Platzman Survey).

53 An IBM Test Scoring Machine was used at Cornell University to evaluate the forecasts made by those taking the elementary meteorology course (Widger and Palmer, 1951).

54 The parts for Taylor and Webb's computer were obtained "by dismantling anti-aircraft fire predictors purchased from war surplus disposals" (p. 4).

55 This despite the fact that this machine was accurate to three figures [Travel Report Wexler to Reichelderfer 20 Dec 46 (Wexler Papers)].

56 For example, Lally (1954) showed that by using deviations from standard values rather than absolute values one could increase accuracy 10-fold in using an analog computer for finding the pressure–height relationship with radiosonde and rawinsonde data.

57 Von Neumann defined the infinite forecast as the atmospheric conditions that prevail when "they have become, due to the lapse of very long time intervals, causally and statistically independent of whatever initial conditions may have existed" (in Smagorinsky, 1983, p. 30).

58 Richard Pfeffer, writing in 1960, said (p. ix), "It is only in the short span of time since the end of World War II that a sufficient quantity of observational data at all levels of the troposphere and the lower stratosphere have become available on a hemispheric basis to define properly even the most gross features of the general circulation."

59 James Holton has written (1979, p. 281), "Progress in numerical modeling of the general circulation has been to some degree dictated by the rate of development in the field of computer technology."

60 In 1951 H.C. Willett wrote (p. 731), "But in spite of all this great expansion of forecasting activity, there has been little or no real progress made during the past forty years in the verification skill of the original basic type of regional forecast, of rain or shine and of warmer or colder on the morrow, the kind of forecasting which first received attention."

Chapter 12

1 *BAMS* **37,** 241 (1956); *BAMS* **57,** 570 (1976).

2 *BAMS* **45,** 337 (1964).

3 *BAMS* **37,** 241 (1956); and *BAMS* **57,** 570 (1976).

4 In 1959 it provided 89% of the funds for research at universities (National Research Council, 1960, p. 6). In 1955 the Federal government spent $22 million on meteorological research, in 1965, $41 million (Bates and Fuller, 1986, p. 179).

5 Beginning in the early 1960s much of the use by airlines of meteorological data became fully automated through the computer programs that prepared flight plans and estimated arrival times (Kraght, 1963), and in 1978 Frederick Shuman wrote (p. 6), "Virtually every commercial flight within or departing U.S. territory flies on a flight plan made from wind and temperature forecasts provided by computer-to-computer links with the National Meteorological Center." One senses the environmental movement of the 1960s in the change of name of the Weather Bureau: in 1965 it became a part of the newly established Environmental Science Services Administration. According to Lorenz (1970a p. 23), "The '60s were the decade when thoughts of weather modification, other than simple cloud seeding, passed from the bizarre to the respectable."

6 The Television Infrared Observation Satellite (TIROS) series was responsible for most of the operational observation by satellite until 1966, when the ESSA (Environmental Science Services Administration) series assumed this role. The Nimbus satellites served mainly to test new instruments and techniques of data collection. The Applications Technology Satellites (ATSs), which had geosynchronous orbits, allowed meteorologists, for the first time, to make "movies" of cloud movements.

7 "Meteorological satellite research program . . ." 24 Oct 58 (Wexler Papers).

8 As we have seen, data push was being felt acutely even before weather satellites. In the 1950s the rate at which weather information arrived at the National Meteorological Center in Washington increased by 5% annually ["Meteorological satellite research program . . ." 24 Oct 58 (Wexler Papers)].

9 Adding to the forecaster's need for rapidity were some of the new weather services, such as warnings of local storms, where predictions had to be made extremely quickly.

10 Proposal for a Study Project 15 Jul 53 (Wexler Papers).

11 This sort of checking was sometimes done by graphical means in the first half of the century (Shaw, 1931, p. 208).

12 The main concern has not been the small errors of measurement, but the large errors that arise in the recording of observations and in the communicating and processing of the records. According to Gustavsson, "Most of these large errors are introduced by manual mistakes or during transmission of data over telecommunication lines without transmission error control." In the mid 1980s when data arrived at the National Climatic Data Center it was subjected to more than 400 checks, "including syntax tests and physical-limits tests, climatological-limit tests, internal-consistency tests, test for time continuity, and cross checks" (Heim, 1988, p. 491).

13 The inverse—if you can simulate it, then you understand it—was not accepted. See the quotation of Smagorinsky in the next section.

14 An indication of the prevalence of statistical analysis in the mid 1960s is provided by the abstracts of 64 talks presented at a conference on micrometeorology sponsored by the American Meteorological Society in 1964 [*BAMS* **45,** 533–550 (1964)]. More than half of the abstracts made reference to statistical analysis of data.

15 Bjerknes names "the isentropic up and down gliding of the air next to atmospheric fronts" and "the effect of sharp anticyclonic curvature of upper-air currents leading to deepening of the downwind trough" as examples of such phenomena (1980, p. 99).

16 *BAMS* **48,** 595 (1967).

17 Station Papa was a weather ship stationed in the Gulf of Alaska, GARP was the Global Atmospheric Research Program (mentioned above), and FGGE was the First GARP Global Experiment.

18 The construction of analog systems, that is, physical models, of atmospheric phenomena has on quite a few occasions been influential. In the 1950s D. Fultz's physical models of atmospheric motion were much discussed; such work by Fultz and others led, according to Smagorinsky (1972, p. 27), to a "flurry of theoretical work." One of Fultz's findings was that there can be more than one stable state for a fluid with a given boundary condition (Charney, 1960, p. 17). Laboratory simulations are still important, as for example the physical model of microbursts devised by P. F. Linden and J. E. Simpson (1985).

19 At that time Greenberg was Director of the Geophysical Research Directorate (cosponsor with the Institute for Advanced Study of this conference) of the Air Force Cambridge Research Center, Air Research and Development Command.

20 Quite a few of the other talks given at this conference reported results of numerical experimentation. See *BAMS* **45,** 533–550 (1964).

21 Thompson here gives several examples of discoveries made by means of numerical experimentation.

22 In the new importance of numerical experimentation, meteorology is typical of the physical sciences generally. The physicist Leo Kadanoff wrote in 1986 (p. 7), "In the last 50 years, a new approach to physics has arisen almost equal in importance to the two "old" branches of theory and experiment. This new type of effort is computational physics: It involves the use of computers large and small to simulate the behavior of physical systems and to work out the consequences of physical ideas, as expressed in mathematical form."

23 The background and complete text of the proposal are given in Smagorinsky (1983).

24 At the Association for Computing Machinery Conference on the History of Scientific and Numeric Computation, held in Princeton in May 1987, John Rice of Purdue University mentioned his own informal study that reached the same conclusion as Lax's.

25 This quotation is from p. 85 of the 1974 English translation. In the preface to the English translation Marchuk wrote, "The present book is devoted to a single, but very urgent, aspect of that problem, i.e., the numerical methods for solving problems of weather prediction and the general circulation of the atmosphere. This aspect is an important one, since effective numerical methods will broaden the domain of possible mathematical formulations of the problems which can be efficiently solved by computers."

26 The following statement by a prominent climatologist, H. H. Lamb, who has long held to a historical, largely descriptive approach rather than a computational, physics-based approach, shows the centrality of computer modeling in modern climatology: "Modelling in a realm as complex, and with as many interactive variables, as the climatic system is primarily an aid to thought and to conceiving the patterns of the real world rather than an automatic provider of accurate or reliable answers. It can suggest probable linkages of cause and effect in the climate system and often the probable order of magnitude of some of the effects. And it is obviously the main way of exploring the possible consequences of human activities . . ." (Lamb, 1982, p. 367).

27 In the 1950s Rossby, seeing the atmosphere as continually interacting chemically with land and water, championed the subject of atmospheric chemistry (Byers, 1960, p. 261). Charney studied biogeophysical feedback mechanisms; he showed that in certain conditions a decrease of vegetation leads to a decrease of rainfall (1980, p. 198).

28 The connection has been particularly strong in Scandinavia: Vilhelm Bjerknes, V. W. Ekman, Bjorn Helland-Hansen, Carl-Gustaf Rossby, and Jacob Bjerknes have all made important contributions to both meteorology and oceanography. The theoretical affinity of the two sciences, together with the diminutive size of oceanography, explains why many oceanographers were trained in meteorology programs. Thus, when World War II caused a sharp increase in the demand for oceanographers, meteorology programs helped to satisfy this demand [*BAMS* **32,** 107 (1951); Haurwitz, 1985, p. 431].

29 This study led to some progress in long-range weather prediction (Reed, 1977, p. 398).

30 He wrote also that meteorology, oceanography, and hydrology would "probably become united at both national and international levels."

Chapter 13

1 ("Nous devons donc envisager l'état présent de l'univers, comme l'effet de son état antérieur, et comme la cause de celui qui va suivre. Une intelligence qui pour un instant donné, connaitraît toutes les forces dont la nature est animée, et la situation respective des êtres qui la composent, si d'ailleurs elle était assez vaste pour soumettre ces données à l'analyse, embrasserait dans la même

formule, les mouvemens des plus grands corps de l'univers et ceux du plus léger atome: rien ne serait incertain pour elle, et l'avenir comme le passé, serait présent à ses yeux.")

2 The sentence was in italics in the original.

3 Dobson presented his paper at a meeting of the Royal Meteorological Society. In the discussion that followed, Richardson praised Dobson's work and said that the failure of his own numerical forecast was apparently due to the phenomenon Dobson had studied—the large effect of small barometric irregularities.

4 (Enfin, nous insisterons beaucoup sur ce point dans la suite, la prévision du temps a des limites . . . l'objectif scientifique le plus important actuellement étant peut-être de préciser, ce qui est prévisible, et ce qui ne l'est pas.)

5 The three equations may be written as follows: $dx/dt = 10(y-x)$, $dy/dt = 28x-y-xz$, and $dz/dt = 8z/3$.

6 A third channel may be claimed: in 1984 Alvin Saperstein, starting from Richardson's model of an arms race, constructed a model that for certain parameter values behaves chaotically (in agreement, according to Saperstein, with real arms races) (pp. 303–305).

7 For a review of the status of research and practice in long-range weather forecasting in the late 1960s, see Namias (1968). For a more recent review, see Namias (1986).

Note on Sources

My main sources have been the research reports, review articles, addresses, and letters published in leading meteorological journals, especially *Quarterly Journal of the Royal Meteorological Society* (abbreviated *QJRMS*), *Bulletin of the American Meteorological Society* (*BAMS*), *Monthly Weather Review*, and *Tellus*. I have made use of the von Neumann Papers and the Wexler Papers at the Library of Congress and the Charney Papers at the Institute Archives of the Massachusetts Institute of Technology, as well as some documents in the archives of the Charles Babbage Institute in Minneapolis and in the library of the National Climatic Data Center in Asheville, North Carolina. Very important sources have been the reminiscences, retrospectives, and historical articles written by meteorologists; the historical articles by George Platzman, Joseph Smagorinsky, and Philip Thompson have been particularly valuable for a participant's reporting of important events, for inclusion of documents that would otherwise be quite inaccessible, and for a meteorologist's perspective on recent history. Correspondence and conversations with meteorologists, those just named as well as Harold Crutcher, Grady McKay, Ray Joiner, and David Ludlum, have been very helpful.

History of meteorology has until recently been a neglected branch of the history of science. There are only a handful of general histories of the science—A.Kh. Khrgian's *Meteorology: A Historical Survey* (1959), Napier Shaw's *Meteorology in History* (1926), H. Frisinger's *The History of Meteorology: To 1800* (1977), W. E. K. Middleton's *Invention of the Meteorological Instruments* (1969), and Karl Schneider-Carius's *Wetterkunde Wetterforschung: Geschichte ihrer Probleme und Erkenntnisse in Dokumenten aus drei Jahrtausenden* (1955)—and these treat the developments of the 20th century in a cursory way if at all. The level of scholarship for 19th-century meteorology has been raised markedly by two recent studies: Gisela Kutzbach's *The Thermal Theory of Cyclones: A History of Meteorological Thought in the Nineteenth Century* (1979) and James R. Fleming's *Meteorology in America, 1800–1870* (1990). The history of the weather services in the United States is ably recounted in Donald Whitnah's *A History of the United States Weather Bureau* (1961) and Charles C. Bates and John F. Fuller's *America's Weather Warriors, 1814–1985* (1986), which deals with the military weather services. Some aspects of 20th-century meteorology are treated in Bernhard Neis's *Fortschritte in der meteorologischen Forschung seit 1900* (1956) and in Robert Marc Friedman's *Appropriating the Weather: Vilhelm Bjerknes and the Construction of a Modern Meteorology* (1989). I have made regular use of biographical memoirs and obituary notices in various journals and of articles from the *Dictionary of Scientific Biography* and from *McGraw-Hill Modern Scientists and Engineers* (where most of the articles are autobiographical).

For the history of computing I have found the following particularly useful: articles in the *Annals for the History of Computing*, Herman Goldstine's *The Computer from Pascal to von Neumann* (1972), and Michael R. Williams's *A History of Computing Technology* (1985). A number of the volumes in the Charles Babbage Institute Reprint Series for the History of Computing have been valuable; see Campbell-Kelly and Williams (1985), Eckert (1940), Hartree (1947, 1949), Horsburgh (1914), and Wilkes *et al.* (1951) in the list of references. The dissertation by Jary Croarken, *The Centralization of Scientific Computation in Britain 1925–1955* (1986), was an important source of information on the scientific use of mechanical calculators.

Sources of information are identified in footnotes throughout the text, but for several chapters I am particularly indebted to one or two sources and acknowledge here: for Chapter 6, Oliver Ashford's *Prophet—or Professor: The Life and Work of Lewis Fry Richardson* (1985) and George Platzman's "A retrospective view of Richardson's book on weather prediction" (1967); for Chapter 9, John M. Stagg's *Forecast for Overlord* (1971); and for Chapter 10, the papers of Charney, von Neumann, and Wexler, and William Aspray's "A new climate for meteorological research and practice" (an early version of Chapter 6 of his 1990 book *John von Neumann and the Origins of Modern Computing*).

The epigraphs for the three parts come from Shaw (1926, p. 8), Bjerknes (1914, p. 14), and Richardson (1922a, p. xi), respectively.

References

Abbe, C. (1878). Short memoirs on meteorological subjects. *Smithson. Inst., Annu. Rep., 1877,* pp. 376–478.

Abbe, C. (1890). "Preparatory Studies of Deductive Methods in Storm and Weather Predictions," Appendix 15 of the Annual Report of the Chief Signal Officer for 1889. U.S. Signal Office, Washington, DC.

Abbe, C. (1890). Untitled report. *Proceedings of the American Association for the Advancement of Science.* **39**, 77.

Abbe, C. (1891). "The Mechanics of the Earth's Atmosphere." Smithsonian Institution, Washington, DC.

Abbe, C. (1907). The progress of science as illustrated by the development of meteorology. *Smithson. Inst., Annu. Rep., 1907,* pp. 287–309.

Abbe, C. (1910). "The Mechanics of the Earth's Atmosphere," 3rd collec. Smithsonian Institution, Washington, DC.

Abercromby, R. (1887). "Weather, a Popular Exposition of the Nature of Weather Changes from Day to Day," Sci. Ser. Vol. 59. Paul, Trench & Co., London.

Air Ministry, England, Meteorological Office (1921). "The Computer's Handbook." His Majesty's Stationery Office, London.

Alter D. (1924). Application of Schuster's periodogram to long rainfall records, beginning 1748. *Mon. Weather Rev.* **52**, 479–483.

Angervo, J. M. (1928). Einige Formeln für die numerische Vorausbestimmung der Lage und Tiefe der Hoch- und Tiefdruchzentra. *Ann. Acad. Sci. Fenn. Ser. A* **28** (10), 1–45.

Archibald, R. C. (1948). "Mathematical Table Makers." Scripta Mathematica, New York.

Aristotle (ca. 340 B.C./1962). "Meteorologica" (with an English translation by H. D. P. Lee). Harvard Univ. Press, Cambridge, MA.

Ashford, O. M. (1985) "Prophet—or Professor: The Life and Work of Lewis Fry Richardson." Adam Hilger, Bristol.

Aspray, W. (1985). Should the term Fifth Generation Computer be banned? *J. Comput. Math. Sci. Teach.,* Spring, pp. 36–38.

Aspray, W. (1987). The mathematical reception of the modern computer: John von Neumann and the Institute for Advanced Study computer. *In* "Studies in the History of Mathematics" (E. R. Phillips, ed.), pp. 166–194. Mathematical Association of America, New York.

Aspray, W. "John von Neumann and the Origins of Modern Computing." MIT Press, Cambridge, MA.

Austrian, G. D. (1982). "Herman Hollerith: Forgotten Giant of Information Processing." Columbia Univ. Press, New York.

Backus, J. (1980). Programming in America in the 1950s—some personal impressions. *In* "A History of Computing in the Twentieth Century." (N. Metropolis, J. Howlett, and G.-C. Rota, eds.), pp. 125–135. Academic Press, New York.

Baldit, A. (1921). "Études élémentaires de météorologie pratique." Gauthiers-Villars, Paris.

Ball, J. (1906). A rapid method of finding the elastic force of aqueous vapour and the relative humidity from dry-bulb thermometer readings. *Q. J. R. Meteoro. Soc.* **32** 47.

Ballard, J. C. (1931). Table for facilitating computation of potential temperature. *Mon. Weather Rev.* **59**, 199–200.

Barger, G. L., ed. (1960) "Climatology at Work: Measurements, Methods and Machines." U.S. Department of Commerce, U.S. Govt. Printing Office, Washington, DC.

Bashe, C. J., Johnson, R., Palmer, H., and Pugh, E. W.. (1986). "IBM's Early Computers." MIT Press, Cambridge, MA.

Basu, J. E. (1984). Jerome Namias, pioneering the science of forecasting. *Weatherwise* **37**, 191–201.

Bates, C. C. (1956). Marine meteorology at the U.S. Navy Hydrographic Office—a resume of the past 125 years and the outlook for the future. *Bull. Am. Meteorol. Soc.* **37**, 519–527.

Bates, C. C., and Fuller, J. F. (1986). "America's Weather Warriors, 1814–1985." Texas A&M Univ. Press, College Station.

Baum, W. A., and Thompson, P. D. (1959). Long-range weather forecasting in the Soviet Union. *Bull. Am. Meteorol. Soc.* **40**, 394–409.

Bedient, H. A., Neilon, J. R. (1961?). "Automatic Production of Meteorological Contour Charts," mimeogr.

Bellamy, J. C. (1949). Objective calculations of divergence, vertical velocity and vorticity. *Bull. Am. Meteorol. Soc.* **30**, 45–49.

Bellamy, J. C. (1952). Automatic processing of geophysical data. *Adv. Geophys.* **1**, 1–43.

Beniger, J. R., and Roby, D. L. 1978). Quantitative graphics in statistics: A brief history. *Am. Stat.* **32**, 1–11.

Benzi, R., and Franchi, P. (1988). Fluid flow, turbulence, and the weather. *Perspec. Comput.* **8** 4–13.

Bergeron, T. (1959). Methods in scientific weather analysis and forecasting. In "The Atmosphere and the Sea in Motion" (B. Bolin, ed.), pp. 440–474. Rockefeller Inst. Press, New York.

Berkofsky, L. (1952). A numerical prediction experiment. *Bull. Am. Meteorol. Soc.* **33**, 271–273.

Berry, F. A., Bollay, E., Jr., and Beers, N. R., eds. (1945). "Handbook of Meteorology." McGraw-Hill, New York.

Besson, Louis (1904) Essai de prévision méthodique du temps. *Annuaire Société Météorologique,* **52**, pp. 92–97.

Bigelow, F. H. (1902). Higher meteorology in the United States Weather Bureau. *In* "Proceedings of the Second Convention of Weather Bureau Officials Held at Milwaukee, Wisconsin, August 27, 28, 29, 1901" (J. Berry and W. F. R. Phillips, eds.), Weather Bur. Bull. No. 31, pp. 19–22. U.S. Govt. Printing Office, Washington, DC.

Bigelow, F. H. (1903). The mechanism of countercurrents of different temperatures in cyclones and anticyclones. *Mon. Weather Rev.* **31**, 72–84.

Bigelow, J. (1980). Computer development at the Institute for Advanced Study. In "A History of Computing in the Twentieth Century" (N. Metropolis, J. Howlett, and G.-C. Rota, eds.), pp. 291–310. Academic Press, New York.

Bigler, S. G. (1981). Radar: A short history. *Weatherwise* **34**, 158–163.

Birkhoff, G. (1980). Computing developments 1935–1955, as seen from Cambridge, U.S.A. *In* "A History of Computing in the Twentieth Century" (N. Metropolis, J. Howlett, and G.-C. Rota, eds.), pp. 21–30. Academic Press, New York.

Bjerknes, J. (1948). Practical application of H. Jeffreys' theory of the general circulation. Programme et Résumé des Mémoires. Réunion d'Oslo, Association de Météorologie, Union de Géodesie et Géophysique International, pp. 13–14.

Bjerknes, J. (1980). Bjerknes, Jacob, Aall Bonnevie. *In* McGraw-Hill Modern Scientists and Engineers, Vol. 1, pp. 98–99. McGraw-Hill, New York.

Bjerknes, V. (1906). "Fields of Force." Columbia Univ. Press, New York.

Bjerknes, V. (1914). Meteorology as an exact science. *Mon. Weather Rev.* **42**, 11–14.

Bjerknes, V., and Sanström, J. W. (1910). "Dynamic Meteorology and Hydrography," Part I. Statics. Carnegie Institution of Washington, Washington, DC.

Bjerknes, V., Hesselberg, T., and Devik, O. (1911). "Dynamic Meteorology and Hydrography," Part II. Kinematics. Carnegie Institution of Washington, Washington, DC.

Bjerknes, V., Bjerknes, J., Solberg, H., and Bergeron, T. "Physikalische Hydrodynamik, met Anwendung auf die dynamische Meteorolgie." Springer-Verlag, Berlin.

Bolin, B. (1952). Studies of the general circulation of the atmosphere. *Adv. Geophys.* **1**, 87–118.

Bolin, B. and Charney, J. G. (1951). Numerical tendency computations from the barotropic vorticity equation. *Tellus* **3**, 248–257.

Boys, C. V. (1895). Scale lines on the logarithmic chart. *Nature (London)* **52**, 272–274.

Brooks, C. F. (1950). Thirty years of the American Meteorological Society. *Bull. Am. Meteorol. Soc.* **31**, 210–214.

Brown, D. P., and Harvey, R. A. (1961). Solar- and sky-radiation integrator. *Bull. Am. Meteorol. Soc.* **42**, 325–332.

Brown, W. V. (1901). A proposed classification and index of weather maps as an aid in weather forecasting. *Monthly Weather Review,* **29**, 547–548.

Brunt, D. (1917). "The Combination of Observations." Cambridge Univ. Press, Cambridge, UK.

Brunt, D. (1933). Editorial. *Q. J. R. Meteorol. Soc.* **59**, 95–96.

Brunt, D. (1934). "Physical and Dynamical Meteorology." Cambridge Univ. Press, Cambridge, UK.

Brunt, D. (1939). "Physical and Dynamical Meteorology," 2nd ed. Cambridge Univ. Press, Cambridge, UK (1st ed. 1934).

Brunt, D. (1944). Progress in meteorology. *Q. J. R. Meteorol. Soc.* **70**, 1–11.

Brunt, D. (1951). A hundred years of meteorology: 1851–1951." *Adv. Sci.* No. 30.

Brunt, D., and Douglas, C. K. M. The modification of the strophic balance for changing pressure distribution, and its effect on rainfall. *Roy. Meteorol. Soc. Mem.* **3**, 29–51.

Buell, C. E. (1961). "Determination of the Feasibility of a 'Winds-Ahead Computer'," Final Rep. Contract No. N189(188)50237A. Naval Weather Research Facility, Naval Air Station, Nofolk, VA.

Burks, A. R., and Burks, A. W. (1988). "The First Electronic Computer, the Atanasoff Story." Univ. of Michigan Press, Ann Arbor.

Burnham, J. C., ed. (1971). "Science in America." Holt, Rinehart & Winston, New York.

Burstyn, H. L. (1971). Theories of winds and ocean currents from the discoveries to the end of the 17th century. *Terrae Incognitae* **3**, 7–31.

Burstyn, H. L. (1984). William Ferrel and American science in the centennial years. *In* "Transformation and Tradition in the Sciences" (E. Mendelsohn, ed.), pp. 337–351. Cambridge Univ. Press, Cambridge, UK.

Buys-Ballot, C. H. D. (1874). "Les changements périodiques de température." Utrecht.

Byers, H. R. (1937). "Synoptic and Aeronautical Meteorology." McGraw-Hill, New York.

Byers, H. R. (1959). Carl-Gustav Rossby, the organizer. *In* "The Atmosphere and the Sea in Motion" (B. Bolin, ed.), Rossby Mem. Vol., pp. 56–59. Rockefeller Institute Press, New York.

Byers, H. R. (1960). Carl-Gustaf Arvid Rossby. *Natl. Acad. Sci., Biogr. Mem.* **34**, 248–270.

Byers, H. R. (1970). Recollections of the war years. *Bull. Am. Meteorol. Soc.* **51**, 214–217.

Campbell-Kelly, M., and Williams, M. R., eds. (1985). "The Moore School Lectures: Theory and Techniques for Design of Electronic Digital Computers," Charles Babbage Institute Reprint Series for the History of Computing, Vol. 9. MIT Press, Cambridge, MA.

Casey, R. S., and Perry, James, eds. (1958). "Punched Cards, Their Application to Science and Industry." 2nd ed. Reinhold, New York (1951).

Cassidy, D. C. (1985). Meteorology in Mannheim: The Palatine Meteorological Society, 1780–1795. *Sudhoffs Arch.* **69**, 8–25.

Chapman, S. (1922). Review of "Weather Prediction by Numerical Process," by Lewis F. Richardson. *Q. J. R. Meteorol. Soc.* **48**, 282–284.

Charney, J. G. (1949). On a physical basis for numerical prediction of large-scale motions in the atmosphere. *J. Meteorol.* **6**, 371–385.

Charney, J. H. (1950), Progress in dynamic meteorology. *Bull. Am. Meteorol. Soc.* **31**, 231–236.

Charney, J. H. (1951). Dynamic forecasting by numerical process. *In* "Compendium of Meteorology" (T. F. Malone, ed.), pp. 470–482. Am. Meteorol. Soc., Boston.

Charney, J. H. (1955). The use of the primitive equations of motion in numerical prediction. *Tellus* **7**, 22–26.

Charney, J. G. (1960). Numerical prediction and the general circulation. *In* "Dynamics of Climate" (R. L. Pfeffer, ed.), pp. 12–17. Pergamon, New York.

Charney, J. H. (1972). Impact of computers on meteorology. *Comput. Phys. Commun.* **3**, Suppl. 117–126.

Charney, J. H. (1980). "Conversations with Jule Charney, a Series of Interviews Conducted by George W. Platzman in 1980." NCAR Tech. Note 298. Massachusetts Institute of Technology, Cambridge, MA; and National Center for Atmospheric Research, Boulder, CO.

Charney, J. H., and Eliassen, A. (1949). A numerical method for predicting the perturbations of the middle latitude westerlies. *Tellus* **1**(2), 38–54.

Charney, J. H., Fjörtoft, R., and von Neumann (1950). Numerical integration of the barotropic vorticity equation. *Tellus* **2**, 237–254.

Chief of Naval Operations (1967). "U.S. Naval Weather Service Computer Products Manual," Doc. NAVAIR 50-1G-522. Department of the Navy, Naval Weather Service, Washington, DC.

Clayton, H. H. (1923). "World Weather." Macmillan, New York.

Coffin, J. H. (1853). On the winds of the Northern Hemisphere. *Smithson. Contrib.* **6**, Article VI.

Coffin, J. H. (1875). The winds of the globe: Or the laws of atmospheric circulation over the surface of the earth. *Smith. Contrib.* **20**.

Conrad, V., and Pollak, L. W. "Methods in Climatology." 2nd ed. Harvard Univ. Press, Cambridge, MA (1st ed. 1944).

Coulson, K. L., Dave, J., and Sekera, Z. (1960). Tables Related to Radiation Emerging from a Planetary Atmosphere with Rayleigh Scattering. University of California Press, Berkeley.

Courant, R., Friedrichs, K., and Lewy, H.(1928). Ueber die partiellen Differenzengleichungen der mathematischen Physik. *Math. Ann.* **100**, 32–74.

Cressman, G. P. (1972). Dynamic weather prediction. *In* "Meteorological Challenges: A History" (D. P. McIntyre, ed.), pp. 179–207, Information Canada, Ottawa.

Croarken, J. (1986). The centralization of scientific Computation in Britain 1925–1955. Ph.D. Dissertation, University of Warwick.

Crutcher, H. L. (1956). Wind aid from wind roses. *Bull. Am. Meteorol. Soc.* **37**, 391–402.

Dady, G. (1967). "Prévoir le temps, évolution des idées et des techniques depuis un siècle." Palais de la Decouverte, Paris.

Dean, M. (1979). "The Royal Air Force and Two World Wars." Cassell, London.

Dedebant, G. (1927). Le champ du déplacement instantané des isobares. *C. R. Acad. des Sci.* **185** (5), 359–361.

Descartes, R. (1637/1965). Les Météores (Discourse on Method, Optics, Geometry, and Meteorology) (Engl. transl. P. J. Olscamp), pp. 263–361. Bobbs-Merrill, Indianapolis, IN.

DeVorkin, D. (1987). Organizing for space research: The V-2 rocket panel. *Hist. Stud. Phys. Biol. Sci.* **18**, Part 1, 1–24.

Dickey, W. W. (1949). Estimating the probability of a large fall in temperature at Washington, DC. *Mon. Weather Rev.* **77**, 67–78.

Dieudonné, J. (1976). Von Neumann, Johann (or John). *Dictionary Sci. Biogr.* **14**, 88–92.

Digges, L. (1576/1975). "A Prognostication Everlastinge Corrected and Augmented by Thomas Digges." London (reprinted in 1975 by Theatrum Orbis Terrarum, Amsterdam, and Walter J. Johnson, Norwood, NJ).

Dingle, A. N., and Young, C. (1965). "Computer Applications in the Atmospheric Sciences." College of Engineering, University of Michigan, Ann Arbor.

Dobson, G. M. B. (1921). Causes of errors in forecasting pressure gradients and wind. *Q. J. R. Meteorol. Soc.* **47**, 261–269.

Doerr, T. N. (1921). Tables of 0.288th powers. *Q. J. R. Meteorol. Soc.* **47**, 196.

Douglas, C. K. M. (1952. Reviews of modern meteorology—4. The evolution of 20th-century forecasting in the British Isles. *Q. J. R. Meteorol. Soc.* **78**, 1–21.

Drazin, P. G., and Tveitereid, M. The problem of turbulence. *Sci. Prog.*(Oxford) **70**, 129–143.
Dunwoody, H. H. C. (1883). "Weather Proverbs." U.S. War Department, Washington, DC.
Eames, C., and Eames R. "A Computer Perspective." Harvard Univ. Press, Cambridge, MA.
Eckert, W. J. (1940/1984). "Punched Card Methods in Scientific Computation." Thomas J. Watson Astronomical Computing Bureau, New York (reprinted, with a new introduction by J. C. McPherson in 1984, by MIT Press and Thomas Publishers).
Eisenhower, D. D. (1948). "Crusade in Europe." Doubleday, New York.
Eliassen, A. (1978). The life and science of Tor Bergeron. *Bull. Am. Meteorol. Soc.* **59**, 387–392.
Elliott, R. D., Olson, C. A., and Strauss, M. A. Analysis of replies to a questionnaire on forecast aids. *Bull. Am. Meteorol. Soc.* **30**, 314–318.
Eriksson, E., and Welander, P. (1956). On a mathematical model of the carbon cycle in nature. *Tellus* **8**, 155–175.
Evesham, H. A. (1986). Origins and development of nomography. *Ann. Hist. Comput.* **8**, 324–333.
Exner, F. M. (1917). "Dynamische Meteorologie." Teubner, Leipzig.
Fassig, O. L. (1907) Guilbert's rules for weather prediction. *Mon. Weather Rev.* **35**, 210–211.
Fedoseev, I. A. (1976). Voeykov, Aleksandr Ivanovich. *Dictionary Sci. Biogr.* **14**, 52–54.
Feldman, T. S. (1983). The history of meteorology, 1750–1800: A study in the quantification of experimental physics. Ph.D. Thesis, University of California, Berkeley.
Feldman, T. S. (1984). Applied mathematics and the quantification of experimental physics: The example of barometric hypsometry. *Hist. Stud. Phys. Sci.* **15**, Part 2, 127–197.
Fjörtoft, R. (1952). On a numerical method of integrating the barotropic vorticity equation. *Tellus* **4**, 179–194.
Fleming, J. R. (1988). Meteorology in America, 1814–1874: Theoretical, observational, and institutional horizons. Ph.D. Dissertation, Princeton University, Princeton, NJ.
Fleming, J. R. (1990). "Meteorology in America, 1800–1870." Johns Hopkins Press, Baltimore, MD.
Friedman, R. M. (1978). Vilhelm Bjerknes and the Bergen school of meteorology, 1918–1923: A study of the economic and military foundations for the transformation of atmospheric science. Ph.D. Dissertation, Johns Hopkins University, Baltimore, MD.
Friedman, R. M. (1982) Constituting the polar front, 1919–1920. *Isis* **73**, 343–362.
Friedman, R. M. (1989). "Appropriating the Weather: Vilhelm Bjerknes and the Construction of a Modern Meteorology." Cornell Univ. Press, Ithaca, NY.
Frisinger, H. H. (1977). "The History of Meteorology: To 1800." Science History Publications, New York.
Fuhrich, J. (1933). Ueber die numerische Ermittlung von Periodizitäten und ihre Beziehungen zum Zufallsgesetz. *Stat. Obzor (Prague),* pp. 471–482.
Fujiwhara, S. (1923). On the mechanism of extratropical cyclones. *Q. J. R. Meteorol. Soc.* **49**, 105–118.
Funkhouser, H. G. (1937). Historical development of the graphical representation of statistical data. *Osiris* **3**, 269–404.
Galtier, C. (1984). "Météorologie populaire dans la France ancienne." Editions Horvath. Le Coteau, France.
Galton, F. (1863). "Meteorographica, or, Methods of Mapping the Weather." Macmillan & Co., London.
Galway, J. G. (1985). J. P. Finley: The first severe storms forecaster. *Bull. Am. Meteorol. Soc.* **66**, 1389–1395, 1506–1510.
Geddes, A. E. M. (1921). "Meteorology: An Introductory Treatise." Blackie & Son, London.
Giao, A. (1929). La mécanique différentielle des fronts et du champ isallobarique. *Mém. Off. Nat. Météorol. Fr.* No. 20, pp. 1–129.
Giao, A. (1935). Rapport sur l'état actuel de la prevision du temps. *In* "Proces-verbaux des séances de l'Association de Météorologie, II Mémoires et discussions." 5th Gen. Assembly, Lisbon, September 1933, Int. Geodet. Geophys. Union, pp. 77–88. Paul Dupont, Paris.

Gifford, F., Jr. (1953). An alignment chart for atmospheric diffusion calculations. *Bull. Am. Meteorol. Soc.* **34**, 101–105.

Gifford, F., Jr. (1972). On the origins of Richardson's rhyme. *Bull. Am. Meteorol. Soc.* **53**, 548.

Gillispie, C. C. (1978). Laplace, Pierre-Simon, Marquis de. *Dictionary Sci. Biogr.* **15**, 273–403.

Gleick, J. (1987a). "Chaos: Making a New Science." Viking Press, New York.

Gleick, J. (1987b) U.S. is lagging on forecasting world weather. *N. Y. Times,* February 15.

Godske, C. L., Bergeron, T., Bjerknes, J., and Bundgaard, R. C. (1957). "Dynamic Meteorology and Weather Forecasting." Am. Meteorol. Soc., Boston, and Carnegie Institution, Washington, DC.

Gold, E. (1923). Review of Albert Baldit's "Etudes élémentaires de météorologie pratique." *Q. J. R. Meteorol. Soc.* **49**, 64–65.

Gold, E. (1930). Obituary notice of F. M. Exner. *Q. J. R. Meteorol. Soc.* **56**, 194–196.

Gold, E. (1934). Incidents in the "March," 1906–1914. *Q. J. R. Meteorol. Soc.* **60**, 121–125.

Gold, E. (1945). Biographical notice of William Napier Shaw. *Obit. Not. Fellows R. Soc.* **5**, 202–230.

Gold, E. (1947). Weather forecasts. *Q. J. R. Meteorol. Soc.* **73**, 151–185.

Gold, E. (1954). Biographical notice of Lewis Fry Richardson. *Obit. Not. Fellows R. Soc.* **9**, 217–235.

Goldstine, H. H. (1972). "The Computer from Pascal to von Neumann." Princeton Univ. Press, Princeton, NJ.

Goldstine, H. H. (1977). "A History of Numerical Analysis from the 16th through the 19th Century." Springer-Verlag, New York.

Goodison, N. (1977). "English Barometers, 1680–1860." Antique Collectors' Club, Woodbridge, Suffolk, NY.

Gray, L. G. (1935). A useful hygrometric calculating device. *Mon. Weather Rev.* **63**, 16–17.

Gregg, W. R. (1935). Progress in international meteorology. *Mon. Weather Rev.* **63**, 339–342.

Gregory, R. (1930). Weather recurrences and weather cycles. *Q. J. R. Meteorol. Soc.* **56**, 103–120.

Gringorten, I. I. (1953). Methods of objective weather forecasting. *Adv. Geophys.* **2**, 57–92.

Grunow, J. (1975). Sprung, Adolf Friedrich Wichard. *Dictionary Sci. Biog.* **12**, 594–596.

Guilbert, G. (1907). Principles of forecasting the weather. *Mon. Weather Rev.* **35**, 211–212 (this is a translation by Oliver Fassig of a manuscript dated 28 September 1905).

Gustavsson, N. (1981). A review of methods for objective analysis. *In* "Dynamic Meteorology: Data Assimilation Methods" (L. Bengtsson, M. Ghil, and E. Kallen, eds.), pp. 17–76. Springer-Verlag, New York.

Guyot, A. (1884). "Tables, Meteorological and Physical." 4th ed. Smithsonian Institution, Washington, DC.

Haas, N. (1924). A method for locating the decimal point in slide-rule computation. *Mon. Weather Rev.* **52**, 29–30.

Hadley, G. (1735/1910). Concerning the cause of the general trade-winds. *Philos. Trans. R. Soc. London.* **39**, 58–62 [reprinted in The mechanics of the earth's atmosphere. *Smithson. Inst., Misc. Coll.* **51**, No. 4 (1910)].

Halley, E. (1686). An historical account of the trade winds and monsoons . . . *Philos. Trans. R. Soc. London* **16**, 153–168.

Halley, E. (1687). A discourse of the rule of the decrease of the height of the mercury in the barometer . . . *Philos. Trans. R. Soc. London* **16**, 104–116.

Hann, J. F. von (1866). Zur Frage über den Ursprung des Föhn. *Z. Oesterr. Ges. Meteorol.* **1**, 257–263.

Hann, J. F. von (1883). "Handbuch der Klimatologie." Von J. Engelhorn, Stuttgart. (Vol. 1 was translated into English by Robert de Courcy Ward, New York, 1903.)

Hann, J. F. von (1901). "Lehrbuch der Meteorologie." Leipzig.

Hann, J. F. von (1906). "Lehrbuch der Meteorologie." 2nd ed. Leipzig.

Hanson, F. V., and Taft, P. H. (1959). Plotting systems for the evaluation of double-theodotie balloon-measured winds. *Bull. Am. Meteorol. Soc.* **40**, 221–224.

Hardy, R., and others. (1982). "The Weather Book." Michael Joseph Ltd., London.

Harrison, H. T. (1951). United Air Lines and its weather service. *Bull. Am. Meteorol. Soc.* **32**, 104–107.

Hartree, D. R. (1947). "Calculating Machines: Recent and Prospective Developments and Their Impact on Mathematical Physics," Charles Babbage Institute Reprint Series for the History of Computing, Part of Vol. 6. MIT Press, Cambridge, MA.

Hartree, D. R. (1949). "Calculating Instruments and Machines," Charles Babbage Institute Reprint Series for the History of Computing, Part of Vol. 6. MIT Press, Cambridge, MA.

Haurwitz, B. (1985). Meteorology in the 20th century: A participants view. *Bull. Am. Meteorol. Soc.* **66**, 282–291, 424–431, 498–504, 628–633.

Heim, R., Jr. (1988). Personal computers, weather observations, and the National Climatic Data Center. *Bull. Am. Meteorol. Soc.* **69**, 490–495.

Hellmann, G. (1895). "Neudrucke von Schriften und Karten über Meteorologie und Erdmagnetismus." Berlin reprinted in 1969 by Kraus Reprint, Wiesbaden.

Helmholtz, H. von (1888/1893). Über atmosphärische Bewegungen. *Sitzungsber. der Köglic Akad. Wiss., Berlin,* pp. 215–228 (an English translation is in Abbe (1891), pp. 78–93).

Heninger, S. K., Jr. (1960). "A Handbook of Renaissance Meteorology." Duke Univ. Press, Durham, NC.

Hertz, H. (1884). Graphische Methode zur Bestimmung der adiabatischen Zustandsgleichungen feuchter Luft. *Meteorol. Z.* **1**, 421–431 (an English translation is in Abbe (1891), pp. 198–211).

Hess, S. L. (1957). A simple analog computer for determination of the Laplacian of a mapped quantity. *Bull. Am. Meteorol. Soc.* **38**, 67–73.

Hewson, E. W. (1963). Engineering meteorology. *Bull. Am. Meteorol. Soc.* **44**, 130.

Hildebrandsson, H. H., and Teisserenc de Bort, L. (1900). "Les bases de la météorologie dynamique," Vol. 2. Gauthier-Villars, Paris.

Hildebrandsson, H. H., and Teisserenc de Bort, L. (1907). "Les bases de la météorologie dynamique," Vol. 1. Gauthier-Villars, Paris.

Hitschfeld, W. F. (1986). The invention of radar meteorology. *Bull. Am. Meteorol. Soc.* **67**, 33–37.

Hoblyn, T. N. (1928). A statistical analysis of the daily observations of the maximum and minimum thermometers at Rothamsted. *Q. J. R. Meteorol. Soc.* **54**, 183–201.

Holton, J. R. (1979). "An Introduction to Dynamic Meteorology," 2nd ed. Academic Press, New York. (1st ed. 1972).

Hooke, R. (1667/1958). Method for making a history of the weather. *In* Thomas Sprat, "The History of the Royal Society." London, 1667 (reprinted, with critical apparatus, by Washington University, St. Louis, MO, 1958, pp. 172–179).

Horsburgh, E. M. ed. (1914/1982). "Handbook of the Napier Tercentenary Celebration or Modern Instruments and Methods of Calculation," Charles Babbage Institute Reprint Series for the History of Computing, Vol. 3. Tomash, Los Angeles.

Hughes, P. (1970). "A Century of Weather Service: A History of the Birth and Growth of the National Weather Service, 1870–1970." Gordon & Breach, New York.

Hughes, P. (1988). FitzRoy the forecaster: Prophet without honor. *Weatherwise* **41**, 200–204.

Humphreys, W. J. (1920). "Physics of the Air." Franklin Institute, Philadelphia.

Humphreys, W. J. (1942). "Ways of the Weather, a Cultural Survey of Meteorology." Jaques Cattell Press, Lancaster, PA.

Institute for Research (1938). "Meteorology as a Career." Institute for Research, Chicago.

Inwards, R. L. (1869/1950). "Weather-lore," 4th ed. Rider, London.

Jacchia, L. G., and Kopal, Z. (1952) Atmospheric oscillations and the temperature profile of the upper atmosphere. *J. Meteorol.* **9**, 13–23.

Jenkins, G. R. (1945). Transmission and plotting of meteorological data. *In* "Handbook of Meteorology" (F. A. Berry, Jr., E. Bollay, and N. R. Beers, eds.), pp. 573–600. McGraw-Hill, New York.

Jewell, R. (1981). The Bergen school of meteorology—the cradle of modern weather-forecasting. *Bull. Am. Meteorol. Soc.* **62**, 824–830 (reprinted from "Research in Norway 1979," pp. 1–8. Oslo).

Johannessen, K. (1975). Sverre Petterssen 1898–1974. *Bull. Am. Meteorol. Soc.* **56**, 892–894.

Johnson, N. (1943). Meteorology and the Royal Air Force. *Q. J. R. Meteorol. Soc.* **69**, 199–205.

Kadanoff, L. P. (1986). Computational physics: Pluses and minuses. *Phys. Today.* July, pp. 7, 9.

Kerr, R. A. (1985). Pity the poor weatherman. *Science* **228**, 704–706.

Khrgian, A. Kh. (1959/1970). "Meteorology: A Historical Survey," 2nd ed. Vol. 1 (translated from the Russian edition published in 1959). Israel Program for Scientific Translations, Jerusalem.

Koelsch, W. A. (1981). Pioneer: The first American doctorate in Meteorology. *Bull. Am. Meteorol. Soc.* **62**, 362–367.

Koeppen, W. (1882). Erlauterungen zur Karte der Haufigkeit und der mittleren Zugstrassen barometrischen Minima Zwischen Felsengebirge und Ural. *Ann. Hydr.,* **10**, 336–344.

Kraght, P. E. (1963. Flight planning with a digital computer. *Bull. Am. Meteorol. Soc.* **44**, 355–363.

Krick, I. P., and Fleming, R. (1954). "Sun, Sea and Sky." Lippincott, Philadelphia.

Kutzbach, G. (1975). Carl-Gustaf Arvid Rossby. *Dictionary Sci. Biog.* **11**, 557–559.

Kutzbach, G. (1979). "The Thermal Theory of Cyclones: A History of Meteorological Thought in the Nineteenth Century." Am. Meteorol. Soc., Boston.

Kwizak, M. (1972). Computer weather forecasting in Canada. *Bull. Am. Meteorol. Soc.,* **53**, 1154–1157.

Lackey, E. E. (1960). A method for assessing hourly temperature probabilities from limited weather records. *Bull. Am. Meteorol. Soc.* **41**, 298–303.

Lally, V. E. (1954). Use of anomalies in the design of a hydrostatis computer. *Bull. Am. Meteorol. Soc.* **35**, 478–480.

Lamb, H. H. (1982). "Climate, History and the Modern World." Methuen, London.

Landsberg, H. (1950). Review of Ratje Mügge's "Fiat review of German science 1939–1946." *Bull. Am. Meteorol. Soc.* **31**, 67–68.

Landsberg, H. (1964). Early stages of climatology in the United States. *Bull. Am. Meteorol. Soc.* **45**, 268–275.

Landsberg, H. (1978). Franz Baur 1887–1977. *Bull. Am. Meteorol. Soc.* **59**, 310–311.

Lansford, H. (1985). To understand the atmosphere. *Weatherwise* **38**, 185–192.

Laplace, P. S. (1814). "Essai philosophique sur les probabilités." Courcier, Paris (reprinted in 1967 by Culture et Civilisation, Brussels).

Lempfert, R. G. K. (1920). "Meteorology." Methuen, London.

Lempfert, R. G. K. (1932). The presentation of meteorological data. *Q. J. R. Meteorol. Soc.* **58**, 91–102.

Linden, P. F., and Simpson, J. E. (1985). Microbursts: A hazard for aircraft. *Nature (London)* **317**, pp. 601–602.

London, J., and Raschke, E. (1983). Fritz Möller 1906–1983. *Bull. Am. Meteorol. Soc.* **64**, 1093.

Loomis, E. (1883). The barometric gradient in great storms. *Am. J. Sci.* **26**, 1–20.

Lorenz, E. N. (1963). Deterministic nonperiodic flow. *J. Atmos. Sci.* **20**, 130–141.

Lorenz, E. N. (1969). Three approaches to atmospheric predictability. *Bull. Am. Meteorol. Soc.* **50**, 345–349.

Lorenz, E. N. (1970a). Forecast for another century of weather progress. "A Century of Weather Progress," pp. 18–24. Am. Meteorol. Soc., Boston.

Lorenz, E. N. (1970b). The nature of the global circulation of the atmosphere: A present view. "The Global Circulation of the Atmosphere" (G. A. Corby, ed.), pp. 3–23. R. Meterol. Soc., London.

Lorenz, E. N. (1983). A history of prevailing ideas about the general circulation of the atmosphere. *Bull. Am. Meteorol. Soc.* **64**, 730–734.

Ludlum, D. M. (1984). "The Weather Factor." Houghton Mifflin, Boston.

MacCready, P. B., Jr. (1957). The Munitalp cloud theodolite. *Bull. Am. Meteorol. Soc.* **38**, 460–469.

Macelwane, J. B. (1952). A survey of meteorological education in the United States and Canada. *Bull. Am. Meteorol. Soc.* **3**, 53–55.

Mader, O. (1909). Ein einfacher harmonischer Analysator mit beliebiger Basis. *Elektrotech. Z.,* No. 36.

Mandelbrot, B. B. (1983). "The Fractal Geometry of Nature," updated and augmented. Freeman, New York.

Marchuk, G. I. (1967/1974). "Numerical Methods in Weather Prediction" (English translation from the original Russian). Academic Press, New York (first published in Leningrad in 1967).

Marggraf, W. A. (1963). "Automatic Data Processing of Weather Satellite Data," Rep. GDA63-0046, Contract AF 19(604)-8861. Geophysics Research Directorate, Air Force Cambridge Research Laboratories. Cambridge, MA.

Mason, B. J. (1970). Future developments in meteorology and their likely impact on weather services and the community. *In* "A Century of Weather Progress" (J. E. Caskey, Jr., ed.), pp. 25–34. Am. Meteorol. Soc., Boston.

Mauchly, K. R. (1984). John Mauchly's early years. *Ann. Hist. Comput.* **6**, 116–138.

McAdie, A. (1917). "The Principles of Aërography." Rand McNally, Chicago.

Merton, R. K., Sills, D. L., Stigler, S. M. (1984). The Kelvin dictum and social science: An excursion into the history of an idea. *J. Hist. Behav. Sci.* **20**, 319–331.

Middleton, W. E. K. (1965). "A History of the Theories of Rain and Other Forms of Precipitation." Oldbourne Book Company, London.

Middleton, W. E. K. (1969). "Invention of the Meteorological Instruments." Johns Hopkins Press, Baltimore, MD.

Miller, E. R. (1933). American pioneers in meteorology. *Mon. Weather Rev.* **61**, 189–193.

Millikan, R. A. (1919). Some scientific aspects of the meteorological work of the United States Army. *Proc. Am. Philos. Soc.* **58**, 133–149.

Millikan, R. A. (1920). Contributions of physical science. *In* "The New World of Science, Its Development During the War" (R. M. Yerkes, ed.), pp. 33–48. Century Company, New York [reprinted in Burnham, (1971)].

Mitchell, C. L., and Wexler, H. (1941). How the daily forecast is made. *In* "Climate and Man." pp. 579–598. Yearbook of Agriculture, U.S. Department of Agriculture, Washington, DC.

Mitchell, J. M. (1986). Helmut Landsberg: Climatologist extraordinary. *Weatherwise* **39**, 254–261.

Moncrieff-Yates, G. B. (1947). A rotary periodograph. *J. Sci. Instru.* **24**, 35.

Moore, W. L. (1898/1899). "Weather Forecasting: Some Facts Historical, Practical, and Theoretical," Bull. No. 25. U.S. Department of Agriculture, Washington, DC (this appeared first in *Forum* of May, 1898).

Moore, W. L. (1914). "Descriptive Meteorology." Appleton, New York.

Mügge, R. (1948). "Meteorology and Physics of the Atmsophere, Fiat Review of German Science 1939–1945." Office of Military Government for Germany, Wiesbaden (the text is in German).

Murgatroyd, R. J. (1972). Upper atmosphere meteorology. *In* "Meteorological Challenges: A History" (D.P. McIntyre, ed.), Information Canada, Ottawa.

Namias, J. (1968). Long range weather forecasting—history, current status and outlook. *Bull. Am. Meteorol. Soc.* **49**, 438–470.

Namias, J. (1986). Autobiography. *In* "Namias Symposium" (J. O. Roads, ed.), pp. 1–59. Scripps Institution of Oceanography, La Jolla, CA.

National Oceanic and Atmospheric Administration (1970). "The National Climatic Center." U.S. Department of Commerce, U.S. Govt. Printing Office, Washington, DC.

National Research Council, Committee on Atmospheric Sciences (1960). "The Status of Research and Manpower in Meteorology," Publ. No. 754. Natl. Acad. Sci.—Natl. Res. Coun., Washington, DC.

National Research Council, Division of Geology and Geography (1918). "Introductory Meteorology." Yale Univ. Press, New Haven, CT.

Neis, B. (1956). "Fortschritte in der meteorologischen Forschung seit 1900." Akad. Verlagsges., Frankfurt.

Neumann, J., and Flohn, H. (1987). Great historical events that were significantly affected by the weather: Part 8. Germany's war on the Soviet Union, 1941–45. I. Long-range weather forecasts for 1941–42 and climatological studies. *Bull. Am. Meteorol. Soc.* **68**, 620–630.

Newton, C. W. (1954). Analysis and data problems in relation to numerical prediction. *Bull. Am. Meteorol. Soc.* **35**, 287–294.

Newton, C. W. (1959). Synoptic comparison of jet stream and Gulf Stream systems. *In* "The Atmosphere and the Sea in Motion" (B. Bolin, ed.), pp. 288–304. Rockefeller Institute Press, New York.

Newton, C. W. (1981a). A fifty-book library and journal selections for a Certified Consulting Meteorologist. *Bull. Am. Meteorol. Soc.* **62**, 71–77.

Nichols, D., ed. (1986). "Ernie's War." Random House, New York.

Panofsky, H. A. (1949). Objective weather map analysis. *J. Meteorol.* **6**, 386–392.

Penn, S. (1948). An objective method for forecasting precipitation amounts from winter coastal storms for Boston. *Mon. Weather Rev.* **76**, 149.

Pernter, J. M. (1903). Methods of forecasting the weather. *Smithson. Inst., Ann. Rep., 1903,* pp. 151–165. [This is a translation of an article in Vorträge zur Verbreitung naturwissenschaflicher Kenntnisse in Wien. Vol. 43, No. 14. The English translation is reprinted also in *Mon. Weather Rev.* **31**, 576–582 (1903)].

Peterson, K. R. (1961). A precipitable water nomogram. *Bull. Am. Meteorol. Soc.* **42**, 119–121.

Petterssen, S. (1939). Some aspects of formation and dissipation of fog. *Geofys. Publ.* **12** (10), 21–22.

Petterssen, S. (1940). "Weather Analysis and Forecasting: A Textbook on Synoptic Meteorology." McGraw-Hill, New York.

Petterssen, S. (1958). "Introduction to Meteorology," 2nd ed. McGraw-Hill, New York (1st ed. 1941).

Pfeffer, R. L., ed. (1960). "Dynamics of Climate." Pergamon, New York.

Phillips, N. A. (1951). A simple three-dimensional model for the study of large-scale extratropical flow patterns. *J. Meteorol. Soc.* **8**, 381–394.

Phillips, N. A. (1966). Review of L. F. Richardson's "Weather Prediction by Numerical Process." *Math. Comput.* **20**, 633.

Phillips, N. A. (1982). Jule Charney's influence on meteorology. *Bull. Am. Meteorol. Soc.* **63**, 492–498.

Phillips, N. A., Blumen, W., and Cote, O. (1960). Numerical weather prediction in the Soviet Union. *Bull. Am. Meteorol. Soc.* **41**, 599–617.

Pihl, M. (1970). Bjerknes, Vilhelm Frimann Koren. *Dictionary Sci. Biogr.* **2**, 167–169.

Platzman, G. W. (1952). Some remarks on high-speed automatic computers and their use in meteorology. *Tellus* **4**, 168–178.

Platzman, G. W. (1967). A retrospective view of Richardson's book on weather prediction. *Bull. Am. Meteorol. Soc.* **48**, 514–550.

Platzman, G. W. (1979). The ENIAC computations of 1950—gateway to numerical weather prediction. *Bull. AM. Meteorol. Soc.* **60**, 302–312.

Pollak, L. W. (1925). Hilfsmittel zur Aufsuchung versteckter Periodizitäten sowie zur harmonischen Analyse überhaupt. *Ann. Hydrogr.* **53**, 209.

Pollak, L. W. (1926). "Rechentafeln zur harmonischen Analyse." J. A. Barth, Leipzig.

Pollak, L. W., and Hanel, A. (1935). Bericht über die numerische Methode von J. Fuhrich zur Ermittlung von Periodizitäten. *Meteorol. Z.* **52**, 330.

Pollak, L. W., and Kaiser, F. (1934). Neue Anwendungen des Lochkarten-Verfahrens in der Geophysik. *Hollerith Nachrichten*, heft 44.

Porter, T. M. (1986). "The Rise of Statistical Thinking, 1820–1900." Princeton Univ. Press, Princeton, NJ.

Poulter, R. M. (1938). Cloud forecasting: The daily use of the tephigram. *Q. J. R. Meteorol. Soc.* **64**, 277.

Poulter, R. M. (1945) Science and weather during the war: A summary of forecasting progress. *Q. J. R. Meteorol. Soc.* **71**, 391–396.

Priestly, C. H. B. (1949). Heat transport and zonal stress between latitudes. *Quart. J. Roy. Meteor. Soc.,* **75**, 28–40.

Pugh, E. W. (1984). "Memories that Shaped an Industry: Decisions Leading to IBM System/360." MIT Press, Cambridge, MA.

Ramanathan, V. (1988). The greenhouse theory of climate change: A test by an inadvertent global experiment. *Science* **240**, 293–299.

Rapoport, A. (1968). Richardson, Lewis Fry. *Int. Encycl. Soc. Sci.* **13**, 513–517.

Reed, R. J. (1977). The development and status of modern weather prediction. *Bull. Am. Meteorol. Soc.* **58**, 390–400.

Reichelderfer, F. W. (1941). The how and why of weather knowledge. *In* "Climate and man," pp. 128–153. Yearbook of Agriculture, U.S. Department of Agriculture, Washington, DC.

Reichelderfer, F.W. (1970). The early years. *Bull. Am. Meteorol. Soc.* **51**, 206–211.

Richardson, L. F. (1908a). A freehand graphic way of determining stream lines and equipotentials. *Philos. Mag.* [6] **15**, 237–269.

Richardson, L. F. (1908b). The lines of flow of water in saturated soils. *Sci. Proc. R. Dublin Soc.* **11**, 295–316.

Richardson, L. F. (1910). The approximate arithmetical solution by finite differences of physical problems involving differential equations, with an application to the stresses in a masonry dam. *Philos. Trans. R. Soc. London, Ser. A* **210**, 307–357.

Richardson, L. F. (1911). The approximate solution of various boundary problems by surface integration combined with freehand graphs. *Proc. Phys. Soc., London* **23**, 75–85.

Richardson, L. F. (1919). Measurement of water in clouds. *Proc. R. Soc. London, Ser. A* **96**, 19–31.

Richardson, L. F. (1922/1965). "Weather Prediction by Numerical Process." Cambridge Univ. Press, Cambridge, UK [the Dover edition (first published in 1965) is an unaltered republication of the original work except that a 6-page introduction (written by Sidney Chapman in 1965) is included].

Richardson, L. F. (1923) Theory of the measurement of wind by shooting spheres upward. *Philos. Trans. R. Soc. London, Ser. A* **223**, 345–382.

Richardson, L. F. (1924a) The aerodynamic resistance spheres, shot upward to measure the wind. *Proc. Phys. Soc., London* **36**, 67–80.

Richardson, L. F. (1924b). Review of E. T. Whittaker and G. Robinson's "The Calculus of Observations, a Treatise on Numerical Mathematics." *Q. J. R. Meteorol. Soc.* **50**, 163–164.

Richardson, L. F. (1925). How to solve differential equations approximately by arithmetic. *Math. Gaz.* pp. 415–421.

Richardson, L. F. (1927). The variation of temperature with height in the first 44 metres above the North Atlantic. *Q. J. R. Meteorol. Soc.* **53**, 295–300.

Richardson, L. F. (1932). The dynamics of wind. *Nature (London)* **129**, 220–221.

Richardson, L. F., Wagner, A., and Dietzius, R. (1922) An observational test of the geostrophic approximation in the stratosphere. *Q. J. R. Meteorol. Soc.* **48**, 328–341.

Riehl, H. (1952). A quantitative method for 24-hour jet-stream prognosis. *J. Meteorol.* **9**, 159–166.

Rigby, M. (1958). Obituary notice of C. E. P. Brooks. *Bull. Am. Meteorol. Soc.* **39**, 40–41.

Ritchie, D. (1986). "The Computer Pioneers." Simon & Schuster, New York.

Robertson, G. W., and Cameron, H. (1952). A planimetric method for measuring the velocity of the Upper Westerlies. *Bull. Am. Meteorol. Soc.* **33**, 387–389.

Robinson, A. H. (1982). "Early Thematic Mapping in the History of Cartography." Univ. of Chicago Press, Chicago.

Robins, A. H., and Wallis, H. M. (1967). Humboldt's map of isothermal lines: A milestone in thematic cartography. *Cartogr. J.* **4**, 119–123.

Rossby, C.-G. (1941). The scientific basis of modern meteorology. *In* "Climate and Man." pp. 599–655. Yearbook of Agriculture, U.S. Department of Agriculture, Washington, DC.

Rossby, C.-G. (1949. On the dispersion of planetary waves in a barotropic atmosphere. *Tellus* **1**, 54–58.

Rossby, C.-G. (1957/1959). Current problems in meteorology. *In* "The Atmosphere and the Sea in Motion" (B. Bolin, ed.), pp. 9–50. Rockefeller Institute Press, New York [this is a translation of "Aktuella meteorologiska problem," published in *Svensk naturvetenskap, 1956.* Swedish Natural Science Research Council, 1957].

Rossby, C.-G., Alvord, C. M., and Smith, R. H. (1929). The tephigram, its theory and practical uses in weather forecasting. *MIT Meteorol. Profession Notes* No. 1, pp. 7–13.

Saltzman, B. (1967). Meteorology: A brief history. *In* "Encyclopedia of Atmospheric Sciences and Astrogeology" (R. W. Fairbridge, ed.), pp. 583–591. Reinhold, New York.

Saltzman, B., ed. (1985). Weather Dynamics. *Adv. Geophys.* **28**, Part B.

Saperstein, A. M. (1984). Chaos—a model for the outbreak of war. *Nature (London)* **309**, 303–305.

Sawyer, J., and Bushby, F. (1951). Note on the numerical integration of the equations of meteorological dynamics. *Tellus* **3** (3), 201–203.

Schnapf, A. (1977). A survey of the United States meteorological satellite program. *Weatherwise* **30**, 180–191, 201.

Schneider-Carius, K. (1955). "Wetterkunde Wetterforschung: Geschichte ihrer Probleme und Erkenntnisse in Dokumenten aus drei Jahrtausenden." Verlag Karl Albert, Freiburg (an English translation, "Weather Science, Weather Research," was published in 1975 for NOAA and NSF by the Indian National Scientific Documentation Centre, new Delhi).

Schumann, T. E. W. (1944). An enquiry into the possibilities and limits of statistical weather forecasting. *Q. J. R. Meteorol. Soc.* **70**, 181–195.

Schuster, A. (1897). On the investigation of hidden periodicities with application to a supposed 26 day period of meteorological phenomena. *Terr. Magn.* **3**, 13–41.

Schuster, A. (1900). The periodogram of magnetic declination . . . *Trans. Cambridge Philos. Soc.* **18**, 107.

Schwerdtfeger, W. (1986). The last two years of Z-W-G [Zentral Wetterdienst Gruppe] (Part 2). *Weather* **43**, 157–161.

Scott, R. H. (1869/1971). On the work of the Meteorological Office, past and present. *In* "The Royal Institution Libriary of Science. The Earth Sciences" (S. K. Runcorn, ed.), Vol. 1, pp. 333–345. Applied Science Publishers, London.

Scott, R. H. (1873/1971). On recent progress in weather knowledge. *In* "The Royal Institution Library of Science. The Earth Sciences" (S. K. Runcorn, ed.), Vol. 1, pp. 378—387. Applied Science Publishers, London.

Scott, R. H. (1883/1971). Weather knowledge in 1883. *In* "The Royal Institution Library of Science. The Earth Sciences" (S. K. Runcorn, ed.), Vol. 2, pp. 67–77. Applied Science Publishers, London.

Sellick, N. P. (1937). A slide rule for reduction of barometric pressure. *Q. J. R. Meteorol. Soc.* **63**, 439–440.

Sen, S. N. (1924). On the distribution of air density over the globe. *Q. J. R. Meteorol. Soc.* **50**, 29.

Shaw, R. H., and Innes, W., eds. (1986). "Some Meteorological Aspects of the D-Day Invasion of Europe." Am. Meteorol. Soc., Boston.

Shaw, W. N. (1903a). On curves representing the paths of air in a special type of traveling storm. *Mon. Weather Rev.* **31**, 318–320.

Shaw, W. N. (1903b). Methods of meteorological investigation. *Mon. Weather Rev.* **31**, 415–420.

Shaw, W. N. (1922). Meteorological theory in practice. *Nature (London)* **110**, 762–765.

Shaw, W. N. (1923). "Forecasting Weather, Second Edition." Constable & Co., London.

Shaw, W. N. (1926). "Manual of Meteorology," Vol. 1 Cambridge Univ. Press, Cambridge, UK.

Shaw, W. N. (1928). "Manual of Meteorology," Vol. 2. Cambridge Univ. Press, Cambridge, UK.

Shaw, W. N. (1930). "Manual of Meteorology," Vol. 3. Cambridge Univ. Press, Cambridge, UK.

Shaw, W. N. (1931). "Manual of Meteorology," Vol. 4 (a revised edition of Part IV, 1919). Cambridge Univ. Press, Cambridge, UK.

Shaw, W. N. (1932). The meteorology of yesterday, to-day and to-morrow. *Sicentia* **51**, 393–404.

Shaw, W. N. (1934). The march of meteorology. *Q. J. R. Meteorol. Soc.* **60**, 101–120.

Sheynin, O. B. (1984). On the history of the statistical method in meteorology. *Arch. Hist. Exact Sci.* **31**, 53–95.

Shuman, F. G. (1963). Dynamical prediction in the mid-troposphere. *Bull. Am. Meteorol. Soc.* **44**, 212–214.

Shuman, F. G. (1978). Numerical weather prediction. *Bull. Meteorol. Soc.* **59**, 5–17.

Simpson, G. C. (1928a). Some studies in terrestrial radiation. *R. Meteorol. Soc. Mem.* **2**, 69–95.

Simpson, G. C. (1928b). Further studies in terrestrial radiation. *R. Meteorol. Soc. Mem.* **3**, 1–26.

Simpson, G. C. (1929). The distribution of terrestrial radiation. *R. Meteorol. Soc. Mem.* **3**, 53–78.

Smargorinsky, J. (1970). Numerical simulation of the global atmosphere. *In* "The Global Circulation of the Atmosphere" (G. A. Corby, ed.), pp. 24–41. R. Meteorol. Soc., London.

Smagorinsky, J. (1972). The general circulation of the atmosphere. *In* "Meteorological Challenges: A History" (D. P. McIntyre, ed.), pp. 3–41. Information Canada, Ottawa.

Smagorinsky, J. (1983). The beginnings of numerical weather prediction and general circulation modeling: Early recollections. *Adv. Geophys.* **25**, 3–37.

Smagorinsky, J., Collins, G. O. (1955). On the numerical prediction of precipitation. *Mon. Weather Rev.* **83**, 53–68.

Smith, W. L., and others. (1986). The meteorological satellite: overview of 25 years of operation. *Science* **231**, 455–462.
Snellman, L. W., and Murphy, A. H. (1979). Man and machine in weather forecasting systems—a symposium. *Bull. Am. Meteorol. Soc.* **60**, 800–803.
Solot, S. B. (1939). Computation of depth of precipitable water in a column of air. *Mon. Weather Rev.* **67**, 100:103.
Spilhaus, A. F. (1950). Progress in meteorological intrumentation, 1920–1950. *Bull. Am. Meteorol. Soc.* **31**, 358–364.
Spilhaus, A. F., and Miller, J. E. (1942). "Workbook in Meteorology." McGraw-Hill, New York.
Staff Members, Institute of Meteorology, University of Stockholm (1952). Preliminary report on the prognostic value of barotropic models in the forecasting of 500 mb height changes. *Tellus* **4**, 21–30.
Staff Members, Institute of Meteorology, University of Stockholm (1954). Results of forecasting with the barotropic model on an electronic computer (BESK). *Tellus* **6**, 139–149.
Staff Members, Joint Numerical Weather Prediction Unit (JNWPU) (1957). One year of operational numerical weather prediction. *Bull. Am. Meteorol. Soc.* **38**, 263–268, 315–328.
Stagg, J. M. (1971). "Forecast for Overlord." Ian Allan, London.
Starr, V. (1980). Starr, Victor Paul. *In* "McGraw-Hill Modern Scientists and Engineers," Vol. 3, pp. 148–149. McGraw-Hill, New York.
Starr, V. P. (1948). An essay on the general circulation of the Earth's atmosphere. *J. Meteor.*, Vol. 5, pp. 39–43.
Stern, N. (1980). John William Mauchly: 1907–1980. *Ann. Hist. Comput.* **2**, 100–103.
Stewart, R. W. (1972). Atmospheric boundary layer. *In* "Meteorological Challenges: A History" (D. P. McIntyre, ed.), 267–281. Information Canada, Ottawa.
Stigler, S. M. (1986). "The History of Statistics." Belknap Press, Cambridge, MA.
Sutcliffe, R. C., and Forsdyke, A. G. (1950). The theory and use of upper air thickness patterns in forecasting. *Q. J. R. Meteorol. Soc.* **76**, 189.
Sutton, O. G. (1949). "Atmospheric Turbulence." Methuen, London.
Sutton, O. G. (1954). The development of meteorology as an exact science. *Nature (London)* **173**, 1112–1114.
Sutton, O. G. (1955). The meteorological office, 1855–1955. *Nature (London)* **175**, 963–965.
Taylor, R. J., and Webb, E. K. (1955). "A Mechanical Computer for Micrometeorological Research," Div. Meteorol. Phys. Tech. Pap. No. 6. Commonwealth Scientific and Industrial Research Organization, Melbourne.
Thompson, J. A. (1922). Meteorology. *In* "The Outline of Science" (J. A. Thompson, ed.), Vol. 3, pp. 761–790. G. P. Putnam's Sons, New York.
Thompson, J. C. (1949). Tables for computing the height of standard pressure surfaces from aircraft reports." *Bull. Am. Meteorol. Soc.* **30**, 286–287.
Thompson, J. C. (1950). A numerical method for forecasting rainfall in the Los Angeles area. *Mon. Weather Rev.* **78**, 113.
Thompson, J. C. (1985). Weather prediction at the local weather bureau office as concepts from the Bergen School came to the U.S. *Bull. Am. Meteorol. Soc.* **66**, 1250–1254.
Thompson, P. D. (1952). "Notes on the Theory of Large-Scale Disturbances in Atmospheric Flow with Applications to Numerical Weather Predictions," Geophys. Res. Pap. No. 16. Geophys. Res. Divi., Air Force Cambridge Res. Cent., Cambridge, MA.
Thompson, P. D. (1957). Uncertainty of initial state as a factor in the predictability of large scale atmospheric flow patterns. *Tellus* **9**, 275–295.
Thompson, P. D. (1961). "Numerical Weather Analysis and Prediction." Macmillan, New York.
Thompson, P. D. (1978a). A history of numerical weather prediction in the United States. *In* "History of Meteorology in the United States: 1776–1976." Am. Meteorol. Soc. Boston.
Thompson, P. D. (1978b). The mathematics of meteorology. *In* "Mathematics Today: Twelve Informal Essays" (L. A. Steen, ed.), pp. 127–152. Springer-Verlag, New York.
Thompson, P. D. (1983). A history of numerical weather prediction in the United States. *Bull. Am. Meteorol. Soc.* **64**, 755–769.

Thompson, P. D. (1984). A review of predictability problem. *In* "Predictability of Fluid Motions" (G. Holloway and B. J. West, eds.), pp. 1–10. Am. Phys., New York.

Thompson, P. D. (1986). "An interview with Philip Thompson" (conducted by William Aspray on 5 December 1986). Charles Babbage Institute, Minneapolis, MN.

Thompson, P. D. (1987). The maturing of the science. *Bull. Am. Meteorol. Soc.* **68**, 631–637.

Tribbia, J. J., and Anthes, R. A. (1987). Scientific basis of modern weather prediction. *Science* **237**, 493–499.

Tropp, H. S. (1982). Mauchly: Unpublished remarks. *Ann. Hist. Comput.* **4**, 245–256.

Tufte, E. R. (1983) "The Visual Display of Quantitative Information." Graphics Press, Chesire, CT.

U.S. Air Force, Cambridge Research Center, Geophysics Research Directorate. (1957). "Exploring the Atmosphere's First Mile," 2 vols. Pergamon, London.

U.S. Congress, House Committee on Science and Technology (1978). "Weather Forecasting—Past, Present and Future." Committee on Science and Technology, U.S. Govt. Printing Office, Washington, DC.

U.S. Department of the Air Force, Air Weather Service (1948). "Machine Methods of Weather Statistics." Air Force Data Control Unit, U.S. Dep. Air Force, Air Weather Serv., Washington, DC.

U.S. Department of the Air Force, Air Weather Service (1949). "Machine Methods of Weather Statistics." U.S. Dep. Air Force, Air Weather Serv., Washington, DC.

U.S. Department of Commerce (1953). "Weather is the Nation's Business." U.S. Department of Commerce, Washington, DC.

U.S. Department of the Navy, Aerology Branch (1953). "Climatology by Machine Methods for Naval Operations, Plans, and Research." U.S. Dep. Navy, Washington, DC.

U.S. Weather Bureau (1963). "History of Climatological Record Forms 1009 and 612-14," Key to Meteorological Records Documentation No. 2.11. U.S. Department of Commerce, Washington, DC.

U.S. Weather Bureau (1966). "Radiosonde Observation Computation Tables." U.S. Govt. Printing Office, Washington, DC.

van Bebber, W. J. (1891). "Die Wettervorhersage." F. Enke, Stuttgart.

Vernon, E. M. (1947). An objective method of forecasting precipitation 24-48 hours in advance at San Francisco, California. *Mon. Weather Rev.* **75**, 211.

Visher, S. S. (1924). "Climatic Laws." Wiley, New York.

von Neumann, J. (1944). "Proposal and Analysis of a New Numerical Method for the Treatment of Hydrodynamical Shock Problems," App. Math. Group Rep. 108.1R AMG-IAS No. 1, Pergamon, New York (reprinted in von Neumann, 1963, pp. 361–379).

von Neumann, J. (1963a). "Collected Works," Vol. 5. Pergamon, New York.

von Neumann, J. (1963b). "Collected Works," Vol. 6. Pergamon, New York.

Wagemann, H. (1932). Brauchbare Methoden zur Vorausberechnung von Wetterkarten. *Ann. Hydrogr. (Berlin)* **60** (4), 136–151.

Walker, G. T. (1925a). On periodicity. *Q. J. R. Meteorol. Soc.* **51**, 337–346.

Walker, G. T. (1925b). Review of the second edition of Exner's "Dynamische Meteorologie." *Q. J. R. Meteorol. Soc.* **51**, 430–431.

Walker, G., and Bliss, E. W. (1926) On correlation coefficients, their calculation and use. *Q. J. R. Meteorol. Soc.* **52**, 73–84.

Webster's Third New International Dictionary, (Philip Grove, Editor-in-Chief), p. 315. Merriam-Webster, Springfield, MA. 1981.

Weedfall, R. O., and Jagodzinski, W. M. (1961). Comments on double-theodolite evaluations. *Bull. Am. Meteorol. Soc.* **42**, 322–324.

Wegener, A. (1911). "Thermodynamik der Atmosphaere." Barth, Leipzig.

Weiss, L. L. (1955). A nomogram based on the theory of extreme values for determining values for various return periods. *Mon. Weather Rev.* **83**, 69–71.

Weston, J. (1945). Meteorlogy behind the wire. *Q. J. R. Meteorol. Soc.* **71**, 424–426.

Whipple, F. J. W. (1917). The significance of the harmonic analysis of diurnal variation of pressure. *Q. J. R. Meteorol. Soc.* **43**, 282–283.

Whipple, F. J. W. (1924a). Review of V. H. Ryd's "Meteorological Problems. I. Traveling Cyclones." *Q. J. R. Meteorol. Soc.* **50**, 81–83.

Whipple, F. J. W. (1924b). The significance of regression equations in the analysis of upper air observations. *Q. J. R. Meteorol. Soc.* **50**, 237–243.

Whitnah, D. R. (1961). "A History of the United States Weather Bureau." Univ. Illinois Press, Urban.

Widger, W. K., and Palmer, T. Y. (1951). A method for using International Business Machines test sheets for student forecasting and verification. *Bull. Am. Meteorol. Soc.* **32**, 298–301.

Wilkes, M. V. (1949). "The E.D.S.A.C.," Report on a Conference on High Speed Automatic Calculating Machines. University Mathematical Laboratory, Cambridge, UK.

Wilkes, M. V. (1985). "Memoirs of a Computer Pioneer." MIT Press, Cambridge, MA.

Wilkes, M. V., Wheeler, D. J., and Gill, S. (1951/1984). "The Preparation of Programs for an Electronic Digital Computer," Charles Babbage Institute Reprint Series for the History of Computing, Vol. 1. Tomash, Los Angeles.

Willett, H. C. (1951). The forecast problem. *In* "Compendium of Meteorology" (T. F. Malone, ed.), pp. 731–746. Am. Meteorol. Soc., Boston.

Williams, A. F. (1950). A "geostrophic map." *Bull. Am. Meteorol. Soc.* **31**, 325–317.

Williams, F. L. (1963). "Matthew Fontaine Maury, Scientist of the Sea." Rutgers Univ. Press, New Brunswick, NJ.

Williams, M. R. (1985). "A History of Computing Technology." Prentice-Hall, Englewood Cliffs, NJ.

Wolff, P. M., and Hubert, W. E. (1964). Selection of computer systems for economical meteorological operations. *Bull. Amer. Meteorol. Soc.* **45**, 640–643.

World Meteorological Organization (1956). "International Geophysical Year 1957–1958, Meteorological programme, General Survey." WMO, Geneva.

Index

International Geophysics Series

EDITED BY

RENATA DMOWSKA
Division of Applied Sciences
Harvard University
Cambridge, Massachusetts

JAMES R. HOLTON
Department of Atmospheric Sciences
University of Washington
Seattle, Washington

Volume 1 BEN GUTENBERG. Physics of the Earth's Interior. 1959*

Volume 2 JOSEPH W. CHAMBERLAIN. Physics of the Aurora and Airglow. 1961*

Volume 3 S. K. RUNCORN (ed.) Continental Drift. 1962*

Volume 4 C. E. JUNGE. Air Chemistry and Radioactivity. 1963*

Volume 5 ROBERT G. FLEAGLE AND JOOST A. BUSINGER. An Introduction to Atmospheric Physics. 1963*

Volume 6 L. DUFOUR AND R. DEFAY. Thermodynamics of Clouds. 1963*

Volume 7 H. U. ROLL. Physics of the Marine Atmosphere. 1965*

Volume 8 RICHARD A. CRAIG. The Upper Atmosphere: Meteorology and Physics. 1965*

Volume 9 WILLIS L. WEBB. Structure of the Stratosphere and Mesosphere. 1966*

Volume 10 MICHELE CAPUTO. The Gravity Field of the Earth from Classical and Modern Methods. 1967*

Volume 11 S. MATSUSHITA AND WALLACE H. CAMPBELL (eds.) Physics of Geomagnetic Phenomena. (In two volumes.) 1967*

Volume 12 K. YA. KONDRATYEV. Radiation in the Atmosphere. 1969*

* Out of print.

Volume 13 E. Palmén and C. W. Newton. Atmospheric Circulation Systems: Their Structure and Physical Interpretation. 1969*

Volume 14 Henry Rishbeth and Owen K. Garriott. Introduction to Ionospheric Physics. 1969*

Volume 15 C. S. Ramage. Monsoon Meteorology. 1971*

Volume 16 James R. Holton. An Introduction to Dynamic Meteorology. 1972*

Volume 17 K. C. Yeh and C. H. Liu. Theory of Ionospheric Waves. 1972*

Volume 18 M. I. Budyko. Climate and Life. 1974*

Volume 19 Melvin E. Stern. Ocean Circulation Physics. 1975

Volume 20 J. A. Jacobs. The Earth's Core. 1975*

Volume 21 David H. Miller. Water at the Surface of the Earth: An Introduction to Ecosystem Hydrodynamics. 1977

Volume 22 Joseph W. Chamberlain. Theory of Planetary Atmospheres: An Introduction to Their Physics and Chemistry. 1978*

Volume 23 James R. Holton. An Introduction to Dynamic Meteorology, Second Edition. 1979*

Volume 24 Arnett S. Dennis. Weather Modification by Cloud Seeding. 1980

Volume 25 Robert G. Fleagle and Joost A. Businger. An Introduction to Atmospheric Physics, Second Edition. 1980

Volume 26 Kuo-Nan Liou. An Introduction to Atmospheric Radiation. 1980

Volume 27 David H. Miller. Energy at the Surface of the Earth: An Introduction to the Energetics of Ecosystems. 1981

Volume 28 Helmut E. Landsberg. The Urban Climate. 1981

Volume 29 M. I. Budyko. The Earth's Climate: Past and Future. 1982*

Volume 30 Adrain E. Gill. Atmosphere–Ocean Dynamics. 1982

Volume 31 Paolo Lanzano. Deformations of an Elastic Earth. 1982*

Volume 32 Ronald T. Merrill and Michael W. McElhinny. The Earth's Magnetic Field: Its History, Origin, and Planetary Perspective. 1983

Volume 33 John S. Lewis and Ronald G. Prinn. Planets and Their Atmospheres: Origin and Evolution. 1983

Volume 34 Rolf Meissner. The Continental Crust: A Geophysical Approach. 1986

Volume 35 M. U. Sagitov, B. Bodri, V. S. Nazarenko, and Kh. G. Tadzhidinov. Lunar Gravimetry. 1986

Volume 36 Joseph W. Chamberlain and Donald M. Hunten. Theory of Planetary Atmospheres: An Introduction to Their Physics and Chemistry, Second Edition. 1987

Volume 37 J. A. Jacobs, The Earth's Core, Second Edition. 1987

Volume 38 J. R. APEL. Principles of Ocean Physics. 1987

Volume 39 MARTIN A. UMAN. The Lightning Discharge. 1987

Volume 40 DAVID G. ANDREWS, JAMES R. HOLTON, AND CONWAY B. LEOVY. Middle Atmosphere Dynamics. 1987

Volume 41 PETER WARNECK. Chemistry of the Natural Atmosphere. 1988

Volume 42 S. PAL ARYA. Introduction to Micrometeorology. 1988

Volume 43 MICHAEL C. KELLEY. The Earth's Ionosphere. 1989

Volume 44 WILLIAM R. COTTON AND RICHARD A. ANTHES. Storm and Cloud Dynamics. 1989

Volume 45 WILLIAM MENKE. Geophysical Data Analysis: Discrete Inverse Theory, Revised Edition. 1989

Volume 46 S. GEORGE PHILANDER. El Niño, La Niña, and the Southern Oscillation. 1990

Volume 47 ROBERT A. BROWN. Fluid Mechanics of the Atmosphere. 1991

Volume 48 JAMES R. HOLTON. An Introduction to Dynamic Meteorology, Third Edition. 1992

Volume 49 ALEXANDER A. KAUFMAN. Geophysical Field Theory and Method.
Part A: Gravitational, Electric, and Magnetic Fields, 1992
Part B: Electromagnetic Fields I. 1994
Part C: Electromagnetic Fields II. 1994

Volume 50 SAMUEL S. BUTCHER, GORDON H. ORIANS, ROBERT J. CHARLSON, AND GORDON V. WOLFE. Global Biogeochemical Cycles. 1992

Volume 51 BRIAN EVANS AND TENG-FONG WONG. Fault Mechanics and Transport Properties in Rock. 1992

Volume 52 ROBERT E. HUFFMAN. Atmospheric Ultraviolet Remote Sensing. 1992

Volume 53 ROBERT A. HOUZE, JR. Cloud Dynamics. 1993

Volume 54 PETER V. HOBBS. Aerosol–Cloud–Climate Interactions. 1993

Volume 55 S. J. GIBOWICZ AND A. KIJKO. An Introduction to Mining Seismology. 1993

Volume 56 DENNIS L. HARTMANN. Global Physical Climatology. 1994

Volume 57 MICHAEL P. RYAN. Magmatic Systems. 1994

Volume 58 THORNE LAY AND TERRY C. WALLACE. Modern Global Seismology. 1995

Volume 59 DANIEL S. WILKS. Statistical Methods in the Atmospheric Sciences. 1995

Volume 60 FREDERIK NEBEKER, Calculating the Weather. 1995